Burkard Wördenweber · Uwe Weissflog

Innovation Cell

Burkard Wördenweber · Uwe Weissflog

Innovation Cell

Agile Teams to Master Disruptive Innovation

 Springer

Professor Dr. Burkard Wördenweber
Global advanced
Visteon Deutschland GmbH
European Corporate Office & Innovation Centre
Visteonstr. 4-10
50170 Kerpen
Germany
bwoerden@visteon.com

Uwe Weissflog
Pathway Guidance - Europe
Rinnengärten 1
34516 Vöhl-Marienhagen
Germany
uwe.weissflog@pathwayguidance.bi

Library of Congress Control Number: 2004118097

ISBN 3-540-23559-0 Springer Berlin Heidelberg New York

Springer is a part of Springer Science + Business Media

springeronline.com

© Springer-Verlag Berlin Heidelberg 2005
Printed in Germany

Typesetting: Digital data supplied by authors
Cover-Design: medionet AG, Berlin
Production: medionet AG, Berlin

Printed on acid-free paper   68/3020 Rw   5 4 3 2 1 0

In Gratitude

To Janet and Gerda

Without their patience, encouragement and continuous support we would not have finished writing this book. For this we are grateful.

Foreword

There are many ways to describe the gap, which a lean company has to jump to become innovative. Some people see the gap between research and design for production, where people with different mindsets find it hard to communicate and work for the same goal. Other people feel that the gap is the schism between effectiveness and efficiency, i.e. trying to do the right thing is not compatible with trying always to doing things right. Other people believe the gap to be caused by the different paradigms of exploitation and exploration. The financial constraints of globally competing companies striving to become more and more lean are leaving fewer and fewer resources for the necessary experimentation to find successful innovations. Whatever the explanation one thing is certain: globally acting companies have to marry short term success with long term sustainability. They have to be at the same time competitive with current products and services and innovative to prepare future products and services.

This book offers a novel view for management to address and implement innovation. It shows that innovation can not be ordered but has to be lived. It illustrates with real life examples, how innovation requires courage to do the right thing and not always just the safe thing. And it shows that courage can be its own reward.

I wish you stimulating reading.

John F. Kill
Senior Vice President – Product Development
Visteon Corporation

Visteon Village, December 2004

Contents

Index of Essays

The essays used in this book describe our personal views and understanding of certain key questions. They are the distillate of experiences we gathered on our journey towards self-organisation and innovation. The essays represent individual viewpoints, and can be read independently from each other. They are connected through the author-reader dialogue of the book, and represent the different facets of innovation found in market and product lifecycles.

Index of Examples

We have used examples to illustrate the validity of the points made in this book. All the examples are derived from the authors' own experience, but names, places and contents have been modified to protect the privacy of the original sources. Any resemblance to other actual cases is purely coincidental. You will recognise examples in the text by their indentation and the accompanying line along the left edge.

Introduction

Our economic environment is characterised by changes that are both rapid and radical. Due to the opening-up of frontiers, and the formation of networks connecting businesses globally, survival does not only require fitness, but also the ability to be innovative. But being fit and being innovative are two quite different things. Although leaders and managers are used to the quest for efficiency and leanness, being innovative lies outside their fund of experience.

This book "Disruptive Innovations and how to master them" pays special attention to the leadership and management of innovation, the development of the mindset required for innovation, and the setting of conditions for self-organisation. It highlights the relationship between a traditional organisation and its complementary innovation teams. The book addresses the manager in his everyday environment, and includes a holistic approach to people, processes and organisations. It provides the information needed to set up innovation cells for rapid development of products, services and technology.

The reader is taken by the hand and led through the process of becoming an innovator, using a step-by-step approach. The pathway is illustrated by numerous examples from industrial experience. The book has a three-layer structure, including an author-reader dialogue, essays and examples. The authors used the process of an innovation cell themselves to write the book.

The book's structure follows a dialogue between an experienced manager of disruptive innovation, symbolised by "●" and a manager or professional who is interested in the formation of successful innovation teams. This person is symbolised by "□" in the main dialogue.

1 Disruptive Innovations – and how to master them

- Do you want to know how to innovate? Do you want to know about successful innovation in rapidly changing environments? How to be successful as a small- or medium-sized organisation amidst the overwhelmingly powerful, big global players?

- □ *Yes - I like your questions, but I have many more on the subject of innovation. I have to admit to being a little confused - tell me about the most important aspects of innovation.*

2 A growing awareness of being stuck

- I have some news for you - you are not alone! You can learn to facilitate innovation, but you will need initiative and courage. Today's business environment is increasingly different to that of the past sixty years. Old methods no longer suffice. Oh - and remember- you have to start with yourself.

□ *Why is today's business climate so special? Don't the leading brands show that success comes from continuing to improve what you have always done? Why should I do the starting - what about those around me?*

Example

Who are you trying to impress?

The struggle for power and influence in a company takes up the time and attention of the management. The customer is neglected, and rival firms are given a further chance. The following example shows how easy it is to overlook the market. Here Bernie, the R&D manager of the company MyCorp was too busy impressing his boss to keep the real situation with the customer in view.

Bernie: The CEO was really pleased with the tech-show we had at EcoStar last week. We rolled out the whole artillery of new designs, which use our revolutionary solar technology. The engineers at EcoStar were left gaping when they saw what we are capable of. I bet you were surprised, Rick.

Rick: It was a pretty impressive show you and your people pulled off, I have to give you that.

Bernie: I hope you will now have an easy time bringing in the orders from EcoStar.

Rick: Well, three months ago that would have been the case. But the trouble is, our competitor SimCorp was at EcoStar just two months ago. They only presented a few new designs, but what they showed impressed EcoStar so much, that SimCorp was awarded the contract for the facelift of EcoStar's flagship product.

Bernie: Couldn't we do anything to prevent SimCorp intruding on our market?

Rick: I tried, but when I approached our head of development, he signalled to me to keep well clear. Let EcoStar see for themselves, he said, how incompetent SimCorp is when it comes to running a big and demanding project with a market leader like EcoStar.

Bernie: How long do you think we will have to wait for EcoStar to realise their mistake?

Rick: In my view we may be waiting in vain. SimCorp has allocated engineers to work at EcoStar. They have taken up residence with the engineers from EcoStar. My hunch is that SimCorp is using the opportunity to learn, and to get themselves into a position where they will be difficult for us to shift.

Impressing a CEO and impressing a customer are two different things. The customer needs to be paid close attention in order to gain confidence in your ability to make innovation happen.

- Wait a moment - one thing at a time. Let's start with a fundamental human paradox:

We like things the way they are, and at the same time we want them to be different! This is one of the causes and also the crux of innovation.

2.1 Collective avoidance

- In order to innovate, you and those around you must become comfortable with change - which means becoming aware of your dependence on safety and the comfort of the familiar. Innovation is the opposite of the avoidance of change. In other words, innovation opens the door to new and exciting worlds, but it also demands that you abandon the safe milieu you have become accustomed to.

☐ *I know I must change, because the world around me has already changed.*

Essay

What is changing?

Most people in today's business environments talk about change. Often the conversation entails many different aspects of change: changing markets, speed of change, changing customer expectations, and so on. Individually taken, none of these seems really unique to our times. But viewed systemically, the impact on how businesses will operate in the future will be - and already is -dramatic.

Example

Be prepared for the unexpected

Looking beyond one's own boundaries is an important part of a successful survival strategy. Sometimes the most advanced solution is already implemented at a place where we least expect it.

In early 2003 Fred had the opportunity to attend his first business meeting in Tallinn, the capital of Estonia. This visit provided him with a powerful experience of how a tiny country, such as Estonia, can outpace a large country such as Germany, even in areas where the larger country views itself as extremely advanced.

When Fred checked into his room he asked if he could borrow an Ethernet cable, and whether DSL access was provided in the rooms. Having access to email and the Internet is a key necessity of Fred's when travelling. The concierge stared back at him in astonishment. The following short, but revealing dialogue unfolded:

Concierge: What do you need a cable for?

Fred: Well, I want to access the Internet and I also want to read my emails. I hope to use DSL because it is faster then a 56K modem.

Concierge: But we are close to one of the first "hot spots" in Tallinn. Just go wireless.

Fred: What is a "hot spot"?

Concierge: Let me explain. When the USSR broke down and we became independent again, there was no cable system, and telephone lines were barely working. Instead of investing in copper wires, the country invested in wireless technology. Today you can access the Internet from any point in Tallinn without a cable. Look for signs, which say "WiFi". There you are within so-called "hot-spots", which are areas where wireless access to the Internet is possible. It is fast, cheap and available in any coffee shop.

> Back home, Fred bought a computer with integrated wireless technology. Whenever he travels within the Baltic States, it is indispensable because of its wireless capabilities. In Germany most of the time he still needs cables.

The manifestations of change are nowhere more visible than in production efficiency and innovation effectiveness. Faster cycle-times, "7x24", "follow-the-sun", and "anywhere and anytime" are business expressions of the changes we experience. Speed in particular has become synonymous with successful business operations. At the same time more people are becoming uncomfortable with their individual part in the situation. Complaints include symptoms of "burn-out" even in young people, helplessness in the face of overwhelming complexity, or even a schism of a person's life, the mind and soul so alienated that they no longer recognize each other.

What lies at the root of change today?

Two developments in the past six decades have been and still are the root forces of the changes we are experiencing today:

- The ability to create a digital extension of the existing physical world, which can exist independently, and which increasingly, we inhabit
- The possibility of unrestricted local and global access

Both are interconnected, co-dependent worlds and are used for both work and pleasure. Their combination provides the basis for major shifts in our socio-economic structures worldwide.

What makes the digital world so significant?

Consider the following: You are the Director of Innovation Monitoring at your company. Your daily tasks include a personal assessment of technology trends in the Far East and North America, the coordination of your innovation scouts working in 12 different countries around the globe, and the analysis of your own company's investments into innovative ideas. This represents a rather normal work situation in many high-tech companies. It is obvious that without your laptop and mobile phone, your company's intranet, the internet and world wide web, SMS, email, WAP, video conferencing, and all the other digital tools and services you take for granted today, you probably could not do your job.

Now consider that the only tools available to you are a fax machine, a host computer, and an ordinary phone. How could you do what is asked of you? Remember- just 15 years ago this was probably all you had at your disposal. Of course you are a good manager and eventually you would have managed to get the job done. But instead of doing it in a day, it may have taken you a month.

The digital world then is very significant, if not revolutionary because it permits:

- Miniaturisation of physical components required for information retrieval and distribution, as well as for human communication, to a point where the components can be carried easily, i.e. even when we travel we can be connected in a physical and digital sense wherever we are.

- Dynamic representation of "actual reality" in symbolic form, so that it can be carried around, shared, manipulated and interpreted easily – anywhere, anytime, and by anyone with appropriate tools and access authority, and

- Creation of one standard of representation of physical reality, which is globally accepted and applied.

The impact of the digital world is comparable to the introduction of a global currency and can be experienced by everyone in daily life. The texting-mania of teenagers, factory automation, or music downloads via the Internet are just a few examples, none of which would be possible without this digitalisation. But another component was necessary to transform the digital world into the global agent for change that it has become: the possibility of almost unrestricted distribution of, and access to its digital content, anywhere and anytime.

What makes unrestricted distribution so significant?

The term "unrestricted distribution" is used here to express the new possibilities people have today of working with each other worldwide because of:

- a technological infrastructure, such as internet, telecommunication networks or GPS, which makes it possible to send information to and receive it from individual people in any part of the world

- a local or national political environment, which allows its citizens free access to this technology, and which restricts interference with the contents of the infrastructure only within its own legal frameworks

- an economic environment which enables increasingly larger numbers of people to afford the tools and services which are necessary to access the infrastructure, and

- a social environment, which fosters learning and utilisation of the various technologies as a normal and ubiquitous part of daily work and life.

When these conditions are met and the technological basis is stable, new forms of work can emerge. This is what is happening today. Consider the exchange of and access to information. Just 10 – 20 years ago it was primarily based on paper; today the exchange happens predominantly in digital form. Financial markets and product development are examples of this trend.

The almost unrestricted movement of people and information worldwide is changing businesses and markets dramatically. Businesses can now move to locations where high quality labour is inexpensive, without fearing fragmentation and isolation of the operation. Products from remote manufacturers can enter local markets, which were previously simply too far away to be considered. We experience this every day, whether it relates to our clothes, which have probably been manufactured somewhere in Asia, or the phone-call from central Europe to a call centre for technical help, which may well be located in Ireland or India.

How do unrestricted distribution and digitalisation affect us?

The combination of the digital world and unrestricted global access and distribution of information is the cause of the move from manual labour to "knowledge work". This transition has direct impact on what we produce, how it is produced, where it is produced, and who is involved in the production. In the case of familiar products we experience the impact directly:

- Digitalisation allows increasing the value of familiar products by adding information content or other features, such as GPS in cars or download features in mobile phones. Often the additions add real value, but sometimes they are of little practical use. But in any case they allow proliferation of the original spectrum of use and extend the evolutionary development of these products.

- Moving manual labour from countries paying high wages to those with low wages, by using factory automation, or both of these, increases availability and affordability of familiar products.

The impact of this movement is that the market differentiation of high-wage countries based on production efficiency is disappearing. Their development is restricted because the structural basis for their manual work is disappearing.

The work left for the labour force in highly developed countries is increasingly in innovative products and services.

What new opportunities may emerge?

Therefore the questions every manager and leader in highly developed countries must ask him- or herself are basically these:

- What will better sustain the growth and competitiveness of my company: extended evolution of a product or product replacements?
- What means exist to create enough new work opportunities in a relatively short time, to counterbalance the drain of existing work towards low-wage/high-skills countries?

Here innovation plays a crucial role, in particular if it is disruptive. Its potential as a creator of new markets -and also new job opportunities- may be the best hope of sustaining or even increasing the high standard of living that countries such as Germany or France have become accustomed to. Opportunities emerge in three new areas:

Digital world and unrestricted distribution

First and foremost, the digital world and unrestricted distribution will develop further. Human ingenuity will provide a more integrated and human world. Many new jobs and opportunities will be in this area. Digitalisation will create, for example, a wave of products that take a physical thing, such as a picture or book and transform it into a digital representation of itself, requiring new competence and skills. Unrestricted distribution will allow storing of such a digital product on any inter- or intranet server, thereby offering it to a much larger community of potential users.

Continuous Innovations

Continuous improvement is a term, which is well known from areas such as industrial production, financial services or transportation logistics. Continuous innovation is a similar process to continuous improvement. Small adjustments and modifications change an existing process or product to make it more innovative. With a series of small changes a process or product is made more and more effective, until it is optimised. The paradigm of continuous innovation is "cheaper, bigger and more combinations of the same". With continuous improvement, a company may maintain cost leadership for some time. In the long-term, continuous improvement is not enough to counteract the forces of global markets. Similarly, continuous innovation may help a company to maintain product leadership for some time. Again, continuous innovation is not a permanent differentiator in globally competing markets.

Disruptive Innovations

The essence of disruptive innovation is the creation of something new. Unlike continuous innovation, where we take what we already have and make it more efficient, (either by adding a new function, or eliminating what is superfluous), the aim of disruptive innovation is to replace what already exists.

In this book we will be particularly concerned with the organisation of disruptive innovation in companies. We will pay attention to the need for effectiveness, rather than efficiency. We will look at the potential conflict an organisation experiences between striving for continuous improvement and the need to address disruptive innovation.

☐ *My immediate motivation for change is that the things that used to work no longer seem so effective. But I also feel a desire to move forward-and that is what change brings. There is joy in change- I want to change!*

● Good! What do you think is involved in changing? Are you prepared even *to make change happen*?

☐ *Hmm...the others don't want change, and I feel insecure about change myself.*

● This is very understandable. I feel like this too. Perhaps you would like to consider that this feeling of insecurity is a fear of facing reality- we prefer to see things as we wish they were, rather than as they are. These days the gap between the two seems so wide. Let's have a closer look at the situation.

2.2 Comfortable versus hungry

- Can you relate to a situation where everything seems to be OK on the surface, but once in a while you get a feeling that something is "wrong"?

☐ *Yes, I know this feeling. But isn't it true that it will get even more un-comfortable if we don't make a change? I'm not sure where the feeling of discomfort comes from...*

- Everyone may feel that hard times are coming. This alone does not mean we will change. If we are not prepared to acknowledge the feelings of doom, there certainly won't be a collective movement. Some of your discomfort may stem from being asked to do more and more things in less and less time- and you see that this does not lead to greater success. You have a feeling you could work differently, but don't have time to check what this means on a daily basis. Nor are your colleagues the types to allow you to discuss such matters.

Example

How the fear of the unknown limits our scope

Imagine the following situation: you are expecting a new technology with interesting new customer benefits. It is still far too early to launch into it. The customer however, can't wait. Would you try to satisfy the customer by launching into the new technology, knowing that you are unlikely to deliver, or would you try to create the new customer benefits with available technology in the hope that the customer is more interested in the benefits than the technology?

What however, if the customer knows neither the new technology nor its potential benefits? Let's listen in on the conversation in the R&D department of MyCorp, which illustrates how the fear of the unknown limits the playing field.

Justin: With the up-and-coming WAVELET we will be able to create jets unlike any that went before.

Fred: I doubt that M^{C}CORP will pay for WAVELETS. You know how they are forcing us to accept lower and lower prices on our conventional jets, and they are a lot cheaper than those you are talking about. And by the way, the WAVELET is not yet qualified for hygiene use.

Justin: Won't we have to offer M^{C}CORP extra value in order to avoid losing further margins on existing products?

Fred: What sort of extra value are you thinking of?

Justin: With the new WAVELETS we will be able to pipe any liquid for the first time!

Fred: And what good is that?

Justin: We can create jets, which wrap around corners and are completely integrated into the command unit.

Fred: If that is the case, then M^{C}CORP will probably want to produce jets themselves and MyCorp will be out of the game.

Justin: I could build a prototype using flex&set pipes to show M^{C}CORP the future functionality.

Fred: You won't do any such thing. I won't invest company money or time on a product which a) the customer won't pay for, b) probably never gains automotive approval, c) creates functions no one asked for, and d) might lead to MyCorp losing the little value add it has in jets. Not whilst I am in charge.

The senior manager was mainly aware of the risks of the new technology. He lacked the trust to venture beyond the present confines of his business. The young engineer was not given a chance to start an innovative project.

2.3 Caught in displacement activities

- Instead of considering change, we occupy our time with "being good". This is what I mean: you have worked in your organisation for years, achieved your chosen career, and have a good life. You do not want to disappoint the expectations of your superiors or the organisation. But deep down you also know that this no longer suffices for your own future or that of the organisation.

☐ *Hang on a minute- I'm confused. What do you mean by "the expectations of the organisation" and "this no longer suffices for the future"?*

Essay

Does progress matter?

It is probably fair to say that the changes we experience today are in part disruptive. Simultaneously these disruptions are destructive and creative; they take opportunities away from us and offer us new ones. They undoubtedly move us forward. They are the manifestations of disruptive progress, which demands disruptive innovation for sustainable success. To go beyond the surface of this, we need to consider the different aspects of disruptiveness including:

- Product and technology
- Organisation and structure
- People, both individually and in groups

Let us turn our attention to these topics one at a time.

Disruptive product and technology

Disruptive products and technologies abound today. The Internet is just one dominant example. It has given rise to a host of services and products unimaginable just ten years ago. Biotechnologies, alternative energies, stealth technology and others provide enough examples, many of which are still in their early development years. What is it that makes these products and technologies so disruptive? For one they allow their users to break away from the mainstream. But this is not all; they also provide their users with an advantage not found elsewhere. Thus they provide strategic advantages for those who invent, produce, and use them.

But the advantage does not come cheaply. The price is high, because the strategic value of disruptive technologies and products only bears fruit if they eventually become the norm. In the long term, every disruptive product or technology will lose the lustre of being disruptive, and something else will take its place. But in its wake, the disruptive product or technology will have opened a new pathway for satisfying human needs. The contribution actually lies in the addition to human experience of new pathways when, but only when, the majority of users accept those products and technologies. When they cross the chasm from technological gimmick to mainstream product, they have served their purpose as disruptors.

If innovative products fail to cross this chasm, they sink back into oblivion; again there are plenty of examples. To cross the chasm, more is needed than mere technological brilliance. The disruptiveness has to be unique, and it has to be transformable into mainstream markets.

For disruptiveness to be authentic and unique, the products or technologies need to be viewed in the larger context of the people who are creating and using them. But this naturally requires a closer look at the organisations and structures in which these people live and work.

Disruptive structure and organisation

The era of the single-handed innovation is a thing of the past. The complexity of the tasks in hand, in particular in disruptive innovation, requires organisations and structures, which enable and foster the creative process. Moreover creation is different from management. The creative process needs its own unique environment. This environment will look very unfamiliar to the manager of today's lean business organisation.

Example

A creative research organisation

External workshops are often used by organisations to give employees a break from their daily routines. What happens during the workshops seldom has lasting impact when the people are back at work. The following example demonstrates how things can be done differently, and how these meetings can become fully integrated into the overall work structure.

Jeffrey is director of ResearchIT, a research and development group at MyCorp. His personal goal was to transform his group into a creative team organisation. He had structured a process, which entailed a sequence of management and "all-hands" group meetings on an annual basis. This process had been part of the group's training curriculum for several years. Each year the training sessions followed a specific theme. In the example described below, the director and Bill his external consultant, designed an "all-hands" meeting, which had "creative collaboration" as the main theme.

Jeffrey: This year I want to focus my group on creative collaboration, both internally and externally i.e. with clients and suppliers. In addition to learning about new ways of working together, I want to find something, which helps my people to remember the key results of the meeting during their daily routine.

Bill: This sounds to me like action learning, a method that can help people gain direct experience of collaboration. Why don't we use small groups and let them create their preferred way of collaboration?

Jeffrey: Yes, this is the right direction, but it needs to be exciting and new. If we use the same process that we used last year, people might get bored.

Bill: Why don't we use enthusiasm and openness as the guiding principles? I know a process where the groups have to answer some core questions in a very unique way.

Jeffrey: OK, let us talk about the questions first- what could they be?

Bill: How about asking the groups to work on questions such as: What is important for you when working with other people?

Jeffrey: I like the approach, but how can we make the answers easy to understand?

Bill: We ask the groups to draw pictures of their answers and not to use words.

Jeffrey: You are joking, this will never work. These are technologists and researchers. Last time they drew something was probably at nursery school.

Bill: What would you say if I told you that many thousand people worldwide are using this approach to foster creativity, many of them in high- tech companies? If you want I can provide a list of references for you to talk to.

Jeffrey: OK, I'll take the risk, but what about creating something that reminds people of their results after the meeting?

Bill: Well, we can take photos of the groups' pictures and of their presentations. Afterwards using this material, we can produce something that reminds people of the event.

With this initial conversation in mind the "all hands" meeting was designed and carried out. Different teams created their versions of creative collaboration. Each group had an appropriate time slot to present its ideas to the other groups. Group presentations followed several different forms - some used acting, others told stories. Presentations and the pictures were photographed and recorded.

Using this material, a small team created a gift for each participant as a token of appreciation. The effect of this activity, although not directly measurable, was significant. The gift became a treasured artefact, which nearly everyone carried around or displayed on his or her desk. Since the group began its quest to become a creative team organisation, it has consistently collected the most scientific awards in its field and is considered by the peer research organisations to be one of the outstanding work places at MyCorp.

These structures for innovation may even be disruptive to the hierarchical organisation, which needs them for its own survival. But it is paramount to the success of disruptive innovation that such structures are implemented. Managers who embark on this path of establishing innovation will encounter the following unfamiliar organisational traits:

- Self-organisation i.e. the realisation that people who are given the appropriate freedom, can organise themselves into success without the traditional management and leadership structures

- Messiness i.e. the observation that the most creative phases of the disruptive innovation process can seem chaotic and without any visible structure

- Non-linearity i.e. decision and creation processes which do not follow the usual linearity of lean processes, but seem unpredictable and fuzzy

These structures may also exist only for a limited time, time enough to get the job done. Then they dissolve, and the people who have been a part of them move to different opportunities. Disruptive structures therefore, are foreign entities in a clean and lean business machine. More like organisms, they spring into existence when needed, and die when their life is over.

Disruptive people

Leaders, managers and professionals of disruptive innovation need to remember: innovation is an act of creation. It requires becoming comfortable with the following:

- Unpredictability and uncertainty during the early stages of an innovation project

- Tolerance towards errors and mistakes which will undoubtedly occur during any innovation

- Patience with and trust in the processes of natural emergence and growth

- The ability to go with the flow

- The ability to withstand the urge to be in control at all times

- Intuition for the "right moment" to intervene, and a sense of timing

These skills are required in addition to all those already required for success in production efficiency. Disruptive innovation is at heart Janus-faced i.e. looking in both directions. It will stir up the status quo only to cross the chasm into the mainstream markets, and become the next ordinary product. People involved in disruptive innovation therefore, need to develop skills, which enable both surfing the waves of innovation, and finding pathways into the orderly world of production.

This cannot be done within hierarchical organisations. Hierarchies are necessary, and are good at maintaining stability and control - they breed conformity in people. They offer shelter, as a harbour protects ships and sailors from the dangers of the open sea. But the purpose of disruptive innovation is 'to go out to sea' and catch the next wave of progress. Just as the harbour master is not the captain of the ship, a different organisation and different people are needed for innovation.

The person steering the ship out into the turbulent waters of disruptive innovation needs courage, intuition and a team of independent, but reliable partners. These are rare treats in today's organisations, which often honour conformity over creativity. Disruptive change provides the opportunity to rediscover these skills, and that is why disruptive change matters so much.

- You can probably recognise certain of your work behaviour patterns that make you successful in your organisation. You have climbed your career ladder, and your boss appreciates your contribution. This is obvious from your annual bonus and the appreciation of colleagues and customers. Here we have a similar situation to that of our childhood - we are reverting to a behaviour pattern, which earned us love and acceptance when we were small. There are many names for this: keeping your clothes clean, sitting pretty, being tidy, not hitting your brother (at least not when mum is watching) etc. But both the child we were and the business adult we have become have a dream of what he or she wants to create. Sometimes, conforming to the expectations of your parents or those of your boss can get in the way of what you and your organisation want to achieve.

☐ *Aha - but isn't this "being good" counterproductive?*

- Yes, but we must ask- for whom or what? It is indeed counterproductive for the sustainable existence of the organisation, but it is very effective for maintaining the status quo of the current situation, and for appeasing authority. It also increases the efficiency of current processes, maintaining the feeling that "all is well", and of course also, that we are safe. Perhaps you have encountered situations, which demonstrate this point.

Example

Dedicated teams can handle high risks projects

Project management is often used for product development. The following example shows where project management may fail. The risk associated with innovation is too large for a loose team structure. Switching over to a full time, dedicated team saved the innovation project for the company MyCorp.

John: That was a silly product feature you promised EcoStar, Bernie. Now that we have to develop solar heaters in earnest we stumble over all the problems, which you haven't solved properly. It just shows that we should never have let you visit the customer on your own.

Bernie: Thanks for the trust. Are you unhappy about the ability of the research department to deliver new technologies, or irritated by the lack of resources in development to tackle a demanding innovation?

John: The second crisis meeting yesterday with EcoStar leaves me angry. I still have no prospect of reaching a technically sound solution for the product feature, which EcoStar thinks we have well established.

Bernie: Maybe I can help you. Why don't we pool resources?

John: We only have twelve months left to the start of production!

Bernie: Great! That is just the challenge an innovation cell needs. I will be happy to put one of my engineers into the innovation cell full-time. There are just two conditions attached: Firstly he gets to work at the location where the action is; secondly, we do this together i.e. any criticism you may have, you will bring first to me.

John: OK. Let 's give it a try. Which one of your engineers can I have?

Bernie: Ben I think. He is a good engineer. Moreover he was involved in setting-up the prototypes for EcoStar in the first place. He will not let you down. When can he start?

John: He can start right this afternoon. And by the way, if anything
         should go wrong, I will give you due warning.

The solar heater was delivered on time. There were two critical situations in the development project, which were mastered because both development and research departments joined forces to resolve them. The customer valued the quick and competent response. EcoStar adopted solar heaters as part of their brand identity. At the time, no competitor of My-Corp was able to master the technical challenge.

- I remember situations where my intervention was in fact required, but I looked the other way in order not to rock the boat. In this way underlying problems can be ignored, so that by the time they loom really large, it may well be too late for change

☐ *You mean we behave like lemmings?*

- Yes, at least partly. Of course as long as you are still on the top of the cliff, you are enjoying the company of your fellow lemmings. Only when you jump off the cliff do you see that this was a bad idea right from the start. The tragedy is that by this stage, when everyone realises what is going on, it is too late - you are already falling. But the good news is that it doesn't have to be like that.

3 Reaching a new perspective

- You can start changing things - and the first step is to begin with yourself.

☐ *But I already feel intimidated by all the changes going on around me. I have an inkling of what changes are necessary, but I feel paralysed because there are so many of them. I also feel afraid- these changes will disrupt my existing environment.*

Example

Common misuse of power

Power is often exerted in order to maintain superiority. The following example shows that even a manager's petty habit e.g. demanding ownership of results in his group, can stifle progress. Bernie, the head of R&D at My-Corp lifts the ban and then watches the innovation cell take off.

Mark: I have a patent application here for you Bernie.

Bernie: Thanks. What do you expect me to do with it?

Mark: Well – now that I am in your innovation cell, I thought I'd have to give you my patents. My boss always checked them, and if he didn't like them, that was the end of it. If he did, then he put his name on top.

Bernie: Do you like your patent application?

Mark: Sure, I wouldn't show you if I didn't.

Bernie: Do the other people in the innovation cell like it?

Mark: Yes, but they felt uncertain as to whether we should show it to anyone.

Bernie: I love to see what you are inventing. But I am even more eager to know that the innovation cell is proud of its own ideas and innovations. Please make sure you write down your own patentable ideas so that the team receives the acknowledgement it deserves. Copy me on the applications, but send them straight to our IP office.

The change of procedure empowered Mark and the whole team in the innovation cell to safeguard their intellectual property. As a result, they all became eager to write their own patent applications, and soon sorted out the previously tedious task of who should benefit from the reward which MyCorp provides for successful patent applications? The inventor would reap the benefit, and Mark became the proud owner of many patents. Bernie enjoyed watching the innovation cell take off.

- These are natural enough reactions. Do you remember Hans Christian Andersen's story of The Emperor's New Clothes? Perhaps you feel like the little boy who cried out "But he has nothing on at all!" You might even consider saying "But I have nothing on at all!"

☐ *Yes, perhaps I could - but this is not the whole problem. I still don't know where to initiate change, or how to carry it out successfully. I don't know how to prepare myself for change.*

- But you have started already, simply by asking the question! Now let's take a closer look at where the need for change actually lies.

3.1 Frame shift: effectiveness versus efficiency

- First of all we have to realise that change comes from a shift in perspective from efficiency to effectiveness.

☐ *Aren't they identical?*

- There is a subtle difference, which one needs to grasp. Here is an example to illustrate it: a good manager is efficient - a good leader effective.

□ *Now I am totally confused.*

- To be efficient is to do things right, whereas to be effective is to do the right things.

Essay

Being successful in disruptive environments

We can see that there are some companies in today's environment that are tremendously successful. They can be found both in the domain of continuous improvement as well that of radical change. EBAY may be seen as an example of the first type, and the Internet itself as an example of the second. In both cases innovation has played a leading role in their success.

What characterises disruptive environments?

The rules for business success in a disruptive environment are different to those we were used to in the past. But how do we know when our environment is actually experiencing disruptive change? How do we know for sure exactly what is going on, before we set out to change what has served us successfully for a long time? The immediate answer is to observe carefully. There are indicators you need to pay attention to, for example:

You may lose a market you thought you owned, and you cannot find the reason in the familiar competitive landscape. Actually your main competitors seem to experience similar losses. Here's an example:

When the strategists of the Boeing Commercial Airplane Corporation were once asked what competition they feared most, the astonishing answer was: the Internet. Rightfully so, because the emergent capabilities for video-conferencing combined with the latent threat of terrorism since 9/11 strengthens the chance that business people will travel less, and use video-conferencing capabilities more often.

You may lose your sometime "category killer" product, although you have added function continuously, while at the same time significantly reducing costs. The latest "category killer" product may have replaced your product without you even noticing. Why use a regular phone and pay high service costs, when Internet telephoning may become mainstream, and free at that?

In these examples, clearly rules have been changed, often because a disruptive innovation had been introduced. So what can you do? You need to innovate to keep a hold on markets, products or technologies.

Exactly what is our dilemma?

Often we know intuitively what is the right thing to do, but for many reasons we are not actually doing it. Maybe the reasons are not clear-cut, but challenge us with dichotomies we cannot easily resolve. Among such dichotomies we often find the following:

We know that by playing it safe, i.e. by conforming to the rules of our existing environment, we strengthen our own position within that system. At the same time we also know that the survival of the whole system, and that includes us, requires stepping outside the system and introducing new rules. Doing so will naturally threaten our individual survival within the existing system.

We have become accustomed to a human centric way of talking about our business systems without following through in the implementations. So we declare teamwork as the highest form of collaboration but still reward or punish individuals. Privately we are aware of our hypocrisies, but in public we ignore them.

And we are afraid. Few people admit to it in public, but the private conversations on the topic are abounding. We fear that by doing the right thing, it will not be right for us individually. So in our worst dreams we see our careers vanish, our material security disappear and the little freedom we have in our private world taken away, because we lack the financial funds that support them.

So what are our options? When we have time to reflect on our situation, we sometimes become aware of the fundamental challenge we face: that progress always requires disrupting the status quo, and that courage is needed to do this.

What is your main role as a manager?

It is necessary to balance production efficiency and innovation effectiveness, not by finding the lowest common denominator, but by allowing each to reach its potential. When we accept this basic dilemma, we may see a way out of it. The basic truth is that both production efficiency and innovation effectiveness are needed for the sustainability of the company. So the question changes, from *which* to choose to *how* to balance the two. It requires us to see that each component, production efficiency and innovation effectiveness, have their natural place in the business life cycle. What may come as a surprise is that they follow and are dependent on each other in a circular fashion.

It is very much like the common example of management vs. leadership: management is symbolised as climbing up a ladder efficiently, even if the ladder is leaning against the wrong wall. Leadership is symbolised as leaning the ladder against the "right" wall before climbing up it. The same analogy is applicable to innovation and production. When we produce a product efficiently, without looking to the left or right, we might end up "climbing the wrong wall". Remember how IBM climbed the Host Computer "ladder" very efficiently and nearly missed the Personal Computer "wall"? But when we have effectively innovated, in particular when the innovation is disruptive, the efficiency of production, marketing and sales is imperative, if we are to cross the chasm into the mainstream markets.

As a manager of innovation, this circular dependency of effective innovation and efficient production needs to be embraced as second nature. A sense of timing and appropriateness is essential. In other words, each approach needs to be developed and practised. An intimate understanding of both approaches will be the basis of your success.

How do you ensure the smooth running of the company?

The value of efficiency is the continuous quest to do the "same" thing better. The word "better" is used here in the business context of improving production processes. Thus with a minimum of tasks, the optimum with respect to time (faster), profit (lower cost, fewer resources), and higher quality can be achieved. Basically we continuously ask: what in the existing system can be changed or eliminated to streamline the system? When we combine the answers to this question with the appropriate feedback and measurement mechanisms (such as Balanced Score Cards), we create a perpetual system. This system reinforces its core intent, which is to become continuously better with each cycle of execution. This may reach a climax at a point where the system simply cannot be stopped, and it becomes a "self-fulfilling prophecy".

When products are improved in this way, the results are perpetual reinforcement of market perception and product functionality. In the mind of the customers they become the premier choice. Brands like Coke, Pampers, Tempo or Melitta represent this situation excellently.

As long as our "ladder" leans against the right wall this is the desired approach. But what if the walls have been changed and we have not recognised it? Then like lemmings, we run faster and faster towards the cliff, from which we will take the plunge into our own business death.

What makes your company innovative?

The value of effectiveness comes from the inherent question contained in it: is our ladder leaning against the right wall? Instead of focusing on "doing things right", the focus of effectiveness is on "doing the right thing". In innovation then, in particular when it leads towards disruptive products and technologies, we go on to create the product most likely to be successful out of a myriad of possibilities. In the creative process, instead of improving continuously what we already have, we engage in an adventure rather like a treasure hunt. On a treasure hunt, many of the parameters that allow us to find the treasure are not apparent at the start. What we have is probably a vision of the treasure itself, such as finding a cure for cancer, or a non-staining clothing material. What we will actually find at the end of the journey is not known when we set out.

Although the paradigm of faster, better, and cheaper is also creeping into the quest for innovation, success or failure turns on additional parameters. Among these are:

- The ability to widen your field of view, before making your choices and diving straight into the details. Such an approach requires patience and perseverance, attributes that do not fit in with the belief commonly held in business that "faster is better".

- The ability to listen to a variety of points of view before synthesising your specific direction. Again, you act in opposition to the popular business belief that specialisation is essential. You need not only to be open to many and diverse aspects, but also to have the competence to understand and interpret a wide-range of views.

- The ability to accept failure as a natural part of the innovative process. This is particularly difficult, because it is counter to the basic intuitive assumption of efficiency, which is: there is definitely a chance of getting it right the first time.

- The willingness to leave behind or give up what works. This may sound like madness to those who have experienced the fruits of success and the joy when everything is running smoothly. But in innovation, especially when it is disruptive, giving up what works may open the door to exciting and even uniquely new possibilities.

By now it has become obvious that the efficient company and the innovative company are probably based on different rules and assumptions. If we want a company to be efficient in production and innovative at the same time, we need to bridge these apparent contradictions. That is exactly what some highly successful companies do today.

☐ *OK- so you are saying we may be doing things as well as we possibly can, but if they are the wrong things, then they won't be doing us any good.*

- That's right, you've got it.

☐ *So how do I find out what the right things are?*

3.2 Looking outwards

- My advice is to sharpen your senses in order to see clearly what is around you. You need clarity not only about your external environment, but also about your internal world. To attain this clarity you need help- it cannot be done alone. This is where you need a network of people who also experience a need for change.

☐ *What do you mean by a network? What makes a network so important?*

Example

Inter-divisional Innovation

In many cases the company divisional structure is there for reasons of product delivery. The core skills are vested in one division, for the sake of efficiency in the delivery of a product or product family. Innovation can take place within one division or, as illustrated in the following case, across divisions. The inter-divisional innovations demand more management attention, but can also be more rewarding, because they represent a new and possibly unique combination of company strengths.

Bill: VacuSump now wants both high and low pressure pumps in the new brushless technology. Trouble is, they don't want to pay more than they already pay for the current low-pressure pump. My department is at its limit with the current technology.

Martin: Is this just the purchasing department pushing for a better price or do the engineers also expect a miracle?

Bill: My engineers tell me that there could be a technical solution. The solution may lie in your department. We could design a pump with an electronic actuator.

Martin: You mean in the very place where it is wettest? That will stretch the technology available in my department.

Bill: Yes, you are right. But I think that is exactly the reason why VacuSump is asking us. We are competent in pumps, and we are competent in electronics. Now we will have to become competent in pumps and electronics together.

Martin: I am happy to give it a try. It would, however, require our departments to cooperate closely. In particular, it would require the two of us to share the ups and downs of the project.

The innovation was implemented successfully. New technology was established between the company division for pumps and electronics. Sadly however, when the division heads changed, the infighting began over who should be taking the most profit on the products.

- Think about the last time you had a tough problem to solve. You had a chance to discuss it with a good friend who also had the competence required to deal with the problem. Remember the conversation. What was it that helped you to solve the problem? During the conversation you often felt safe to talk very openly without fear of being misunderstood. It was a conversation based on trust. This means you need to connect with people you trust.

Example

The need for path finding

Technical idealism often drives expectation. A technically attractive solution, for example, can cloud the view of a development engineer to such an extent that he leaves commercial considerations to one side. On a larger scale, an attractive new technology can create a surge in the industry. This despite the fact that a commercially viable path to the new technology is not obvious, and may even prove to be illusive. The following example shows that serious path finding is necessary even when dealing with trends, which are widely agreed upon.

Martin: We must start to develop high-voltage technology for all of our lighting products. I have just been to the big SE conference. Every speaker worth his salt is now propagating the new high-voltage technology electrical architecture. We can't afford to be at the old standard with our products.

Bernd: It will be an expensive move.

Martin: We will have to set aside a company budget for the changeover.

Bernd: I would be happy to move to the new architecture, if I could see some benefit in it for the consumer. Are you sure that our customer can pass the expense of the change over to the end consumer?

Martin: I can't answer that for all our customers. But I think I might just know the type of consumer who would be prepared to pay extra...

Two years later the technology was introduced for the more affluent owners of pick-up vans. These appreciated the extra acceleration for traffic-light starts, and also used the van as an electricity generator for outdoor activities. The company used the very specific consumer requirements as a lever to launch the new technology, and became market leader.

☐ *But people at work or in my neighbourhood all have commitments and so never have time for this kind of conversation.*

● Perhaps "having no time" is a synonym for "not wanting to". Don't waste your time on people who don't want change. Fortunately you have other choices. For example, you could look beyond your immediate circle. You will find that the less people depend on you, or are under obligation to you, the more open is the communication between you.

Essay

What is it that the successful really do?

First and foremost the successful are open-minded and curious. They have the ability to recognise unfamiliar patterns in a sea of complexity and ignorance. They have a sense for needs and desires not yet obvious to others. Nokia and its approach to mobile communication in the Nineties is a good example of such capabilities. Then of course, successful people are capable of filling the gap between their insights and the emerging needs of the market, providing appropriate products and services. But how do they do this? A lot of the behaviour of successful people is nothing to do with technology; in fact it seems to be the cultivation of the following human traits, which make a difference:

- Sharpened senses, which lead to clarity about what is going on within and around them

- Awareness of their own blind-spots, automatic habits and addictions

- Values beyond the material

- Courage to face the unknown and unpleasant, combined with the ability to take risks

- A personal way of enthusing and engaging others in a common goal or vision

Successful people also value themselves - not in an egocentric way, but valuing themselves as a living being, in whom the new and unfamiliar can flourish until others can see and appreciate it.

The successful sharpen their senses

The way the successful use their senses is geared towards obtaining clarity and understanding. They see, hear, feel and intuit with a clarity unfamiliar to most of us. To attain this they constantly need to "rid" their senses of myriads of external stimuli, which bombard them continuously and with ever-increasing speed. To do so they foster and in part re-learn core human skills such as listening, being patient, focusing, being compassionate, and developing a true interest in other people. The latter is of the utmost importance. Many of the skills that need to be developed require the help of others. To train listening skills one needs someone to listen to. And "listening" goes beyond just hearing what comes out of the other person's mouth. The kind of listening that allows the person listened to, to be truly heard, requires an open mind and the deployment of the whole body. This kind of listening is actually a very intense activity, demanding the active presence of the whole person. We will probably remember this sort of conversation, even if it took place long ago, because we felt truly heard.

Using our newly sharpened senses is hard work. There are so many things that distract us from the essential and important. But here too, others can help us. If we are truly interested in the development of other people, maybe someone else is truly interested in our development. Building a network of such people is like building a continuously developing organism. To reach such a state it is necessary to attain a degree of clarity about oneself. This is a huge challenge in itself, because it confronts us with our blind spots and deficiencies, and our very self-confidence may be called into question.

The successful are aware of their habits and addictions

Most of the time we act out habits that we have acquired over a long period of time. It is as if we were on "automatic pilot". Among other reasons, habits are developed when we experience that a particular behaviour pattern achieves a personal goal or desire. If we repeat certain behaviour patterns always creating the same results, we will eventually repeat them unconsciously. This is a good thing in many situations, because it is the basis of efficiency. But it is important that the intended result still matches the situation where the behaviour is applied. What should we do then, when we have forgotten the original reasons for our habits, but still continue to act them out, despite their inappropriateness in a given situation?

When do habits become addictive?

If we repeat our habits even when they damage or hurt us, we may have become addicted to them. We can easily relate personal experiences to this causal chain. Our lifestyles include perhaps smoking habits, alcohol abuse and eating disorders, which are all examples of addictions we may have. But where is the relationship to innovation, or more relevantly, which habits may inhibit innovation? And when do these work-habits become addictive? Let us look at the following potential candidates:

- Playing it safe
- Egotism
- Focus on material gains

These three may not represent a complete list, but they are representative of an increasing inability to approach new horizons. Essentially they point to the core question: what am I willing to let go? A closer examination may provide some answers.

When we *"play it safe"*, we often exchange what we feel is right for what gives us personal safety. We conform to the rules of an organisation despite the fact that we know that they make no sense. Our gain is that the system keeps us within it, therefore providing us with the income we need for our personal life. In hierarchical structures, we tend to praise the people above us and treat badly the people below us. We assume that we will get protection from above, but need not bother about those below. Effectively we sell our souls to satisfy our basic needs. This is a devilish deal indeed, because it damages our creativity and ingenuity. Slowly we become people who just sit and wait for retirement, assuming that then at least, we will experience real life.

The *egocentric* approach neglects the fact that we need others in order to be fully alive. This is meant not in the sense of using people, but in the awareness that we are all equals in the duality we face: we want to be our individual selves, and we also want to be an integrated member of the community. By focusing exclusively on our own individual needs and gains, we poison our workplaces with spite, jealousy, and anger. We experience the results of this daily, from common mobbing, to violence and even killings at the work place. In this way we lose our ability to build effective communities. This is a tragedy, because as part of a community, we would enjoy the safety we so desperately desire.

The *focus on material gains* is effectively an imbalance in our value system. It is an age-old problem, found particularly in western business approaches. What appears to be new is the intensity and precision with which it is applied. Thus only that which can be measured, counted or otherwise rationalised is seen as valuable. Balanced Score Cards abound, but so far there are none existing that enable motivation, foster reliability, or create trust and understanding.

The work-habits mentioned above have a lasting and significant impact on the underlying assumptions about the ways we organise businesses and see our co-workers. Firstly, they restrict our view, making us increasingly blind to new opportunities. Next they foster the assumption that opportunities are dwindling. A scarcity mentality results. Together these send us down a descending spiral into a mood of doom and gloom, selfishness, and isolation - what a future!

Personal change – or how to overcome habits that are no longer appropriate

In the course of time, we often forget the reasons and original situations, which led to our automatic responses. This means that problems arise when we act in an automatic way, because the present situation does not warrant such behaviour. The successful can question their own behaviour and adjust it according to the situation, without calling their identity into question. They have kept or redeveloped the ability to switch the "autopilot" off, and steer their life on "manual" when needed. Using mindfulness and awareness they can bring long-forgotten reasons for habits to the fore, analyse the appropriateness to the present situation, and shed them if they no longer seem useful. In contrast, people who can and will not go through this constant inquiry may become so dependent on their habits that these develop into addictions. This is true for our private life as well as in business.

Only when we are aware of and mindful of ourselves, others and the world around us, do we see the many possibilities available to us. People who want to innovate must start with the assumption of unlimited resources, in particular when it concerns disruptive innovation.

The successful rediscover values beyond the material

The successful already possess or develop the ability to strike a balance between two extreme mentalities – scarcity and abundance. They do this with the unwavering aim of getting close to their own reality and that of their environment. One approach successful people use is to broaden and balance their value system beyond purely material values. For example, trying truly to care for people. We have all come across people who do this, and can probably recall how pleasant it was to be in their presence. Successful people also know that value systems (except for some core values relevant to human interactions), need re-evaluating once in a while. This does not mean that successful people give up their values readily, but that they examine carefully their value system according to their lifecycle position and the context in which they operate.

In simple words, successful people need to be passionate about the "truth" - not in the sense of an absolute truth, but in an ongoing personal struggle to match their own beliefs and assumptions with their life experiences. They are highly motivated by narrowing the gap between what is actually happening, and their interpretation of or desires for it. In the course of this quest they develop a rather natural way of living in our world.

Striving to become whole again

During this process, successful people may not only gain balance, but also become aware of the many ways life can be experienced. This opens up their senses to the possibilities for innovation and progress. The more they see, the more abundant the possibilities become. Thus the act of rediscovering values beyond the material is primarily something that enhances their personal potential for success, including material gains. Balance and awareness become essential skills for a sustainable good life.

The dichotomy of having and being

There is one other aspect of successful people that deserves mention here. Erich Fromm wrote about it in his book "To Have or To Be?" (1976). The distinction between the modes of having and being is something the successful seem to understand very well. But it is not intellectual understanding that counts here; it is the ability to transform understanding into action. For example, when a good listener listens to us, we experience them as someone who truly wants to understand what we say. We feel fully appreciated. This allows us to engage together in open and meaningful conversation.

It is hard to travel this path of "being" in these modern times of ours. The environment is basically focused on "having", which means that we are only appreciated if we conform to the "material value" flavour of the day. Children experience this daily when they go to school – not wearing the "right" brand may cause anything from simple teasing to complete isolation. The successful are able to live with this dichotomy. They are steadfast in their personal quest for "being", and still do not condemn others for their "having" modes.

Taking risks and being courageous

Life is difficult, but when we accept this fact, the difficulty seems to vanish. The successful have embraced this basic truth. They do not try to evade the reality around them; rather they embrace it, with all the conflicts it entails. In business they are aware of the irresolvable tension of striving to uncover the *"truth"*, maintaining their *position* within the communities they are a part of, and recognising their desire for *harmony* in these communities. The only safe place in this triangle of opposing forces is a constant striving to get close to the "truth". This cannot be achieved in absolute terms. Every time a decision to "tell it as it is" needs to be made, careful evaluation of this triangle of forces must be carried out. If it is decided that harmony in the group at a specific point in time is the most important thing, they may still find ways of telling the "truth", despite its negative impact on the harmony of the group. The same is true for their position in the power hierarchy. They know that this striving for the truth is the only chance of a sustainable future for the group. Any violation of this principle can create irrecoverable damage to themselves and other group members.

The courage of successful individuals allows them to act like this. Courage does not mean the absence of fear. Rather, that despite fear of losing their position in a group or disturbing harmony, they move forward, acting according to what seems most appropriate to maintain momentum in a certain situation. This is done despite the fact that they may not be the official wielders of power. If they are in power, they proceed, keeping in mind the progress of the group and the dignity of its members.

Wants and capabilities

To attain to all of this requires more than just the wish to attain it. Without a thorough examination of one's life, the personal growth required for a leader will not just happen. This is an ancient realisation, which Socrates already stated: "the unaware life is not worth living". How does one develop the skills needed to examine one's own life? It seems that taking risks, caring for others, and a human centric value system are key factors for success on the path. There is an amazing by-product - people who engage in this quest radiate some sort of energy to the people around them, creating a field of positive influence.

It seems that this field enables us to be open to the new and unexpected.

The Yin and Yang of efficiency and effectiveness

When we bring this back into the domain of innovation and progress, it becomes obvious that here we are facing similar challenges. Every group faces a dilemma if engaged in innovation within an organisation as well as striving for continuous improvement of its existing processes. Creating something new may often be experienced as "throwing a wrench in a smoothly running machine". Those who are responsible for this smooth running will not appreciate it. What then, are the options that satisfy the necessity of efficiency in daily operations as well as effectiveness of innovations?

We propose to separate these two - efficiency and effectiveness - into the different modes of operation they truly are: the mode of managing, and the mode of creation. They cannot be combined into one form of organisation because they are structurally completely different. Applying the rules of managing to the world of creating, or vice versa, is like playing chess by following the rules of poker - it is unlikely to produce the desired results.

What makes these two so different? Research on the topic is abounding; here we refer primarily to the work of Humberto Maturana and the late Francesco Varela.

There is however a relationship between the two modes of efficiency and effectiveness. When we use the analogy of the ancient Yin and Yang symbol, we can see a way out of the dilemma. The creative mode always precedes the managing mode. When an idea has reached the stage where it can be transformed into a product, this transformation will eventually lead to the question of how such a product can be efficiently produced, marketed and serviced. If we think this through, we also need to acknowledge that every product will one day have served its purpose, and that it will disappear from the markets. So the cycle has to start again. In any organisation that strives for sustainability, the innovative and the managing mode depend on each other unceasingly. When we realise that one precedes the other, we have a way of successfully dealing with this dichotomy.

What then, is required to master the situation? A bold step needs to be taken. Much like the broadening of the individual's value system and the awareness of his habits, we need to see reality for what it is: efficiency and effectiveness are different! Because of these differences they require separate approaches to organisation, processes, modes of thinking and acting, skills of the people involved and technologies applied.

This is tough. It is tough because the prevailing principles of our business world are those of efficiency, as we know them from production and organisational experience. They can be applied to all aspects of the life cycle, whether organisational or product-related. The principles of efficiency are such a strong paradigm because the underlying structure fosters belief that everything can be streamlined to flow without disruption. This is compounded by the belief that only certain people can organise and lead, while others are merely capable of following. Leading and organising are also considered to be superior skills when compared to simple task of following. Tragically, this chain of thought has led to a basic assumption, in particular in organisations, that leaders and managers are more valuable than those they lead and organise. This "value" is expressed through their knowledge, their financial wealth and their social status. The reality that also needs to be acknowledged, is that unfortunately with respect to creativity and innovation, this way of thinking is rather counter-productive.

Successful Innovators enable others to engage

What choice then, is available to someone interested in effective innovation, in particular when it is disruptive innovation? Awareness of the structural differences of creating and managing, and of the existing imbalances in the belief systems will help, but is not enough. Insight alone does not create change. But there are several options which have been tried in practice and which seem to work. Among them are the following:

- Develop a sense for appropriate timing of innovation

- Select able and enthusiastic followers

- Develop and communicate enticing and compelling goals ("vision of the future")

- Have the courage to make the first step, even when standing alone

Developing a sense for appropriate timing requires the openness and broadness described in previous paragraphs. All the senses are needed to spot the point in time that might be just right for investing in disruptive innovation. This happens through laborious work, which cannot be done alone. A network of "scouts" is required, their ears tuned to everything that happens in a particular environment that could influence the project. This includes many aspects, from research to after-sales support. Like a big array of antennae, this network might be able to detect even the slightest variation pointing to opportunities and needs for the innovation project.

Selecting able and enthusiastic followers requires two things: an understanding of the network of skills required for the project, and the ability to communicate the project goal so that others can understand, engage in and become enthusiastic about it. Only with the emotional involvement of the person initiating the project can all this be achieved.

It takes courage to move forward, despite the many voices predicting failure and disaster. Having the courage to take the first step, even on one's own, requires believing in one's own intuition and knowledge. Having a group of people around who are as enthusiastic as oneself helps to overcome doubts. It is useful to train for this task. When past experience provides positive feedback, one develops more confidence in the approach. The amazing experience of those who have followed this path is that the more often they trust themselves, the more often success is lurking just round the corner.

- Sometimes the courage to expose how dependencies preclude good communication, helps connect people on a new level. This could even work with those who have "no time".

□ *But isn't that dangerous?!*

- You've got the gist of it! The Chinese symbol for crisis combines a sign for "danger" with one for "hidden opportunity". Where you have one you are likely to find the other. Unless you are prepared to tell others what really moves you, you will never find out what really moves them. You simply must leave your comfort zone!

Example

The power of risk assessment for Innovation projects

Risk is a form of probability. Risk in innovation exists because of the unknown nature of the innovation project. For many innovation projects it may never be possible to measure the probability of success in advance. In this case, forecasts on a subjective basis may be the only approach leading to an assessment of risk. The following example illustrates the power of risk assessment for a successful implementation of an innovation project.

Bernie: In our project portfolio we have one heavyweight project, EureCar, which we can neither bring to a short-term success nor terminate without loss of face. Any suggestions as to its future?

Russ: There is a lot attached to EureCar. MyCorp is heading the European EureCar project and all our customers subscribe to it. Our top management has high expectations and they have allocated a big budget of 16 million € for it. I am counting on this project to keep me in good pay until my retirement.

Bernie: I ought to be happy about the opportunity, but the chances of getting a product out to pay me, after you have retired, look slim. The controller is painting a rosy picture, when he shows a return on investment of 200% in 10 years. When I multiply the probabilities for a successful introduction of technology (25%), extension of regulations (30%), product placement (50%) and customer acceptance (25%) I am down to a probability of less than 1%.

Russ: Well, it's simple then. We only have to find an innovative product that removes one of the risks, and that hopefully can pay for itself.

Sam: How about Dure-Car. We have been thinking about Dure-Car on and off, but never found a good reason to launch it. It will introduce the new technology in a simple form and does not need new regulations. If we sell it to the customers as pre- EureCar, then they might just lap it up.

The controllers had rated Dure-Car with a return on investment of about 2%. It had never so much as reached top management attention. Now it was introduced as the precursor to EureCar. The 1 million € for Dure-Car was well invested. Its association with EureCar raised its popularity amongst the customers. In the end its return on investment grew to over 40%. More importantly it broke down the technology barrier, paving the way for EureCar.

3.3 Looking for the new and unexpected

- Moreover, you will have to search for the novel and the different. By this I mean that you must prepare yourself mentally for the unexpected. It's rather like a perfectly reflecting mirror - everything can be reflected in it. But if the mirror is tarnished, it only reflects a fraction of what is really there.

□ *What if I don't find anything new - or I find what I hoped not to find?*

- There is an important distinction to make. If you start off with firm expectations, you narrow down your possibilities. You must be prepared to go empty-handed in order to stumble across the unexpected.

Essay

Can you learn disruptive Innovation?

There are times when life seems to be happening elsewhere. We may feel that good things only happen to others. Even worse, we may even think that we ourselves are part of the problem. We don't want to admit to ourselves that we are stuck. When we have reached the point where we realise how helpless we are, then it is time to reconsider and 're-frame'.

Are you happy with your future?

Optimists tend to act according to their positive idea of the future, which makes them happy people. Pessimists are more likely to act according to reality, which makes them down-to-earth and rational. Pessimists are much better at evaluating the possibilities of failure, whereas optimists tend to create reality from their dream. However, the beauty of it is that optimists, as irrational as they are, are more likely to succeed, because they take chances no rational person would dream of taking. Innovation can be viewed as a future need taking shape as a product. The innovator then, is the optimist who is motivated by the opportunity to make new things into reality. A lot of research has gone into what causes motivation for innovation. The good news from the researchers is that the skills of taking chances appropriately, and resilience in the face of defeat, are not inborn traits. They can be acquired.

Feeling right, but somehow unsuccessful?

Let's look first at the acquisition of the opposite of these attitudes or skills. The feeling of helplessness, for example, is acquired over a long period of time, characteristically in the following three steps: not being in control, losing hope and learning to be helpless.

Not being in control

The research done by Martin Seligman demonstrates that the feeling of helplessness starts with the feeling of "not being in control".

When Seligman was an undergraduate studying psychology, the head of his department was engaged in experiments with dogs – experiments that were failing miserably. A group of dogs were given mild electric shocks through the grid on which they were standing, while simultaneously hearing a tone. The aim of the experiment was to see if the dogs could be conditioned to leap over a low wall to a place of safety, if they heard the tone alone, without being given a shock. Strangely, all the dogs just lay down, even though they were being continuously subjected to shocks.

Seligman contemplated the idea that the dogs, having been subjected to *random* shocks, had become helpless. They had no control over what was happening to them, so they just gave up and lay down. Running the experiment again, he gave the dogs the ability to stop the shocks by pushing a panel with their noses. This produced results. The dogs that had a means to control what was happening to them took evasive action by leaping over the barrier.

Lack of control reduces motivation in humans as well. Not being in control is the first step towards helplessness.

Losing hope

We experience fear when we are not in control and the situation around us remains unpredictable. If this predicament prevails for a long stretch of time, our view of it also undergoes a strange transformation. We invent an explanation of why we are in this particular situation. In other words we have adapted to the situation. We start to defend our explanation that actually puts us in the right. This process is known as 'framing'. It is the accommodation to a situation we cannot alter, the frame of which is set up by someone else who is in control.

Learning to be helpless

But what happens when the situation changes, and we are provided with the opportunity once again to take control of our lives? Researchers were surprised to find that when the external situation alters, permitting change, people do not automatically reframe. The explanation created to account for being in the particular circumstance probably provided a sense of purpose. To give this up, even though it is false, becomes doubly difficult. To adapt to the new situation becomes impossible. It is difficult to see new perspectives, or realise that there are now new opportunities. Thus they feel as though they are in the right, but are still failing.

Learned helplessness is something you are also likely to find at work. Employees will put up with the situation within and surrounding the company. For example, they will adapt their own effort to the minimum required. By not being a "swot" they avoid the envy or anger of their colleagues. Thus employees slowly adapt to boundaries they were set, or which they set themselves. Usually they feel safer this way. They may not notice however, that they are missing opportunities to raise their influence or broaden their horizons. The company loses an important part of its energy, and the employees inadvertently relinquish an important part of their motivation.

Do you want to be master of your own success?

The process of learned helplessness can be reversed. We may even be able to learn to become optimists. We can certainly reframe. Let's start by looking at our views concerning our situation.

Permanence, Pervasiveness, and Personalisation

First of all we have to address our experience of having control, and change our explanation for our misfortune. There are three dimensions to explanatory style: Permanence, Pervasiveness, and Personalisation.

Permanence means here, whether one believes that something will always be there and cannot ever be overcome. It is concerned with time and duration.

Pervasiveness has to do with extent. If you fail at a job, does that mean that you are entirely worthless, or simply that you failed in one area and need to try something different. There are people who begin to fail at everything just because their marriage fails. They see one failure in a universal context.

Personalisation is about whether you blame yourself, other people or circumstances for your failure. This is the trickiest of the three, because there is a sense in which it promotes irresponsibility. There are times when we are responsible for a failure, and we need to admit it.

Learning optimism

It is actually possible to help pessimistic people become a lot more optimistic. People, who felt doomed to a life of failure, have come to experience and even enjoy success. It requires that they realise that a pessimistic attitude has a direct effect on their actions, and that their attitude is self-defeating. But this can be changed! It isn't so much a matter of avoiding adversity, but how we choose to respond to it.

Example

Technology looking for a product

Sometimes a new product is looking for the right technology. At other times the new technology needs to find a first product. This was the case with the following example:

Jason: You asked me to research piezo technology. How do you think we should limit the scope of work for piezo?

Bernie: Well, we know that ceramic pipes can transmit ultra-sound. We also know that so far we haven't got very good results transmitting ultra-sound in ceramic pipes. If we don't want to overtax ourselves, we should first look at functions that require low levels of ultra-sound.

Jason: I'm pretty sure powder dispensers require only low levels of ultra-sound.

Bernie: How about creating one or two demonstrators for piezo, which show the technology and the application in powder dispensers. That way we can approach our customers, and shamelessly exploit their expertise to find the right product.

MediPlus was the first customer to get enthusiastic about piezo. They looked at the technology demonstrators with keen interest and came back a few weeks later with an engineering request to develop a miniature pipette. Once established, the technology was later used in other, completely unrelated products.

Reframing

Reframing then becomes a method we can learn to apply. We just have to choose an opportunity that allows us to venture beyond our usual limits. Once we have addressed our fear, we may even be able to enjoy the opportunities provided by changes and reorganisations in companies.

Example

The challenge of a contradiction

Some say that engineers have two modes of operation: one as a problem-solver, the other as a project manager. Most engineering training goes into problem solving. But unless the engineer is taxed by a problem, he will probably remain in the mode for which he received the least training, namely that of project manager. As a generalisation this statement is not true. However, it should be noted that many well-trained engineers really come to life when they are faced with a seemingly intractable problem. In the following example Hugh rose to the challenge when faced with the request to make the micro-pump smaller rather than bigger:

Hugh: This is the third prototype of a micro-pump. The first one was made of plastic and it became too hot. The second one, made of metal, was still too hot. The third one has cooling fins but it still suffers thermal problems. We will just have to make the micro-pump bigger.

Bernie: What does SplendiCorp say?

Hugh: The SplendiCorp engineer has just checked the surrounding area for the micro-pump. He has lowered our temperature limit even further.

Bernie: Can he accommodate a bigger micro-pump?

Hugh: I didn't dare ask. Ideally he would like it smaller, of course.

Bernie: What would a micro-pump look like, if I asked you to make one half the size of your last prototype?

Hugh: I don't think it could be done.

Two weeks later Hugh proudly presented his fourth prototype. It was not only smaller, but it was also made of plastic. Hugh had surpassed himself again. No wonder that SplendiCorp was delighted.

Are you prepared to go where there is no right and wrong?

When we leave the state of learned helplessness, we also have to leave its associated certainty of "being in the right". If not, we are likely to miss opportunities that lie outside our old frame of reference. New opportunities will first appear without prior warning. We will recognise them by the fact that they are neither right nor wrong.

If we have experience of getting stuck once before, we may be reluctant to accept a new opportunity. Before going forward and accepting a new challenge, we may just want to double check whether or not the new situation will give us good prospects and influence. If we draw up the portfolio of motivation, we can identify the following desirable, yet different options:

1. options with good prospects and influence – these will be our most favoured opportunities

2. options with good prospects and a little influence – these may have to be endured for a while

3. options with meagre prospects but plenty of influence – these will carry us forward for some time

but options with poor prospects or influence should be avoided.

Do you have to do everything yourself? and who guides you?

Rarely can we undergo the process of successfully reframing on our own. The change happens most naturally in groups, for example, in the family or a team. Within the group, everyone has to find his or her freedom, as well as their limits. The boundaries of the group provide safe space within, and help the individual to overcome fear. The group itself can foster a spirit of adventure, and lead each individual to exceed his/her own limitations. So choose your group carefully, and you will find that it will lead you to new shores.

□ *Is this the core of innovation? It sounds like a gamble to me. Is that what you really want?*

4 Taking the plunge into the unknown

- Hang on a minute- let's get this straight. You wanted to know how to innovate. So you should ask yourself the following questions: "Am I prepared to deal with the unexpected?", "Am I ready to choose an unknown path?", "What gives me the courage to take the first step?"

Essay

Go with the flow – the courage to take the plunge

You have probably experienced frustration when plans and ideas, which were enthusiastically developed, are not implemented. Similarly, you may have encountered similar situations in your private life. The holiday you were looking forward to so much is postponed for a reason which hindsight reveals as unimportant. Your presence in the audience at your children's play has to be cancelled at the last moment, because your boss needed the next PowerPoint presentation. The list of unfulfilled promises and expectations is endless, only matched by your accumulated frustration and guilt. The same is often true for the ideas and changes we want to realise in our workplaces and organisations. Business plans end their life without implementation on many a dusty shelf. The breakthroughs crushed by the all-too-familiar "don't rock my boat", which comes in the disguise of rational arguments such as:

- Great idea, but we tried this before and it failed

- Wonderful, but at the moment we really do not have the time nor the resources

- Super, but is this really relevant to our business?

The list can be extended endlessly. There are many ways of dressing up a "no" with apparently authentic arguments, especially when risk is involved. If you want to start something new and unfamiliar, the message from others often seems to be: you are alone. Officially we support and encourage the new, but when it leads to disrupting the status quo, support vanishes and encouragement easily fades away. Change is OK, but only if the existing system stays intact, and "I'm not affected". Exceptions exist, but in most cases they all start at the very top of an organisation. Grassroots movements are discouraged. An enterprising spirit at the middle and bottom level is expected, but not supported if not in accordance with top management.

This has led to a peculiar situation. Effectively all attempts at change, in particular for disruptive innovation, depend on the initiative of the leaders at the top of the organisation. The rest of the organisation lies dormant or paralysed. This flies directly in the face of the call for networked organisations, distributed processes and innovation. Leaders and followers have an equal part in the situation. Followers take fewer risks, because they fear punishment from the levels above. The top levels become convinced that only they can actually introduce change, because they see no action from the levels below. Increasingly, everything has to flow from the top. But in a networked and distributed environment, the top management has no means of actually knowing what's going on, if it is relying on hierarchically organised processes. Many attempts at change fail because of this, but sometimes an idea is so strong that it prevails despite the obstacles.

Often the middle management in many companies want to make a difference. They want to contribute to their companies' success. The middle and lower echelons of organisations are in the best position to sense precisely the pulse of their environments and markets. When it comes to innovation, this is a hopeful area. Leaders and managers of innovation projects can count on the following conditions, which may foster the need for innovation today:

- Many companies have squeezed out most of the potential for efficiency gains by applying the same methods: lean operations, automation and work-force reductions. Often this has led to similar productivity gains in a rather large group of firms. This has at least two consequences: efficiency as a market differentiator is slowly vanishing, and profit margins are moving towards zero.

- The move from high wage to low wage countries is a last resort to cut costs. Industries, such as Information Technology, Automotive or Textiles seem to benefit considerably from this strategy.

- The current trend towards novelty inspired and fuelled largely by digital technologies. New and fresh is "in" and whoever comes up with the right products to satisfy this hunger can develop real market leadership.

- The interconnectedness of product and services into marketable lifestyle systems. Mobile communications, wireless computing, and music on demand are examples demonstrating the power of this trend.

In particular the high-wage countries will increasingly have to look to innovation to keep their work forces employed. There is really little or no choice here. It may look as if the working world can be split into the domains of low-wage/efficient production and high-wage/effective innovation. But this thinking only fools us. Neither do low wages necessarily mean low levels of education. In fact in an increasing number of low-wage countries, the standards of education are equal if not higher than in Germany. Herein lies a real danger. When the conditions described above no longer represent localised phenomena, but combine and join forces, they pose a real threat to high-wage countries.

Top management and leadership may be in control of investment money, and may wield power over resources, but they still depend on the insights of people closer to the markets and potential customers. This provides opportunities for innovation managers. If they can find the right words to portray to executive management the challenges stated above, and simultaneously have the courage to lever the creative energy of the whole organisation, they empower themselves to initiate processes leading to new opportunities.

Example

Successful product strategy in 2 days

Speed, timing and immediacy are key elements in developing and executing strategic plans. Imagine you have a very complex software development organisation with about 700 developers, spread across multiple international locations. You are at the edge of a new wave of industry needs. Your plan must reflect these trends and needs, and it must be created in such a way that all key developers are engaged in its implementation right from the start.

You are responsible for creating the plan, and know from past experience that traditional approaches have not been very successful. Here is what a middle manager at MyCorp did:

He convinced the leadership team at MyCorp to use Open Space, a workshop design for large groups, based on self-organisation. This was an unfamiliar concept for the company. The executive team had strong concerns and doubted the result of such an approach. Because of this they involved themselves in the preparation of the meeting. The CEO and his staff familiarised themselves with the process, and asked the questions they wanted answered during the meeting.

For 2 days, ninety people from sales, marketing, product development and product support met and talked. Professionals, managers and executives were included. At the end of the second day, a 200-page document had been produced, covering 18 strategic questions from market analysis to product innovation.

At the beginning of the first day, the middle manager that had instigated the workshop, explained the principles of the process: self-organisation, responsibility for oneself, commitment and freedom of choice. Then the whole group collected the questions it wanted addressed during the meeting. These complemented those questions framed by the executive team. An agenda emerged, and each question was assigned a time and location during the 2 days.

Participants then signed-up for topics to which they could either contribute or from which they could learn. After that the groups started their work in multiple parallel sessions. Each group documented its own key-findings, and presented them in plenary meetings, organised between the group sessions. The facilitator collected the session results whenever a group session report was made, gradually creating the final document while the workshop was still in progress. At the end of the second day, a total of 18 group sessions had been held and their results had been recorded electronically.

In a final closing session all results were acknowledged, and everybody had a chance to say something about the significance of this meeting for his or her work. Follow-up meetings were held afterwards, ensuring customer feedback and detailed development and implementation plans.

Because of this initial success, MyCorp frequently used unorthodox, large-group meetings to update and complement its strategic product development plan. The developers were able to implement these plans without any time lag. Clarity about the task at hand, and enthusiastic engagement of participants, were key ingredients because all those who needed to be involved had participated from the beginning. Executives and management adjusted to the unfamiliar form of self-organisation because it gave them good results. The bottom line was impressive: this approach had significant impact on moving the company from $80million to $450million revenue in just 4 years.

But initiating processes leading to innovation is not going to be easy for you. You are still a part of an established system requiring you to follow its rules. However the increasing demand for innovative products, in particular those that disrupt the status quo, leaves some room to manoeuvre. To be successful requires some careful preparation, including an understanding of what to "pack" and what to "leave behind" on the journey towards innovation.

What do you need to let go of?

The most challenging skill you need to develop is that of accepting your fears. This is particularly threatening to our understanding of ourselves, because officially, we are fearless. We believe that courage is the absence of fear, although it is exactly the opposite. Courage is something we may have despite our fears. By facing and overcoming fear, we become increasingly courageous. When we are engaged in the creative act of bringing innovation to life in its most natural form, we face the fear of the unknown. What we have forgotten, largely because of our dependency on the efficiency paradigm, is how to do this. Very often small things open the door to new territory. Instead of being very precise and detailed, we may develop the courage to be less precise, and allow room for human ingenuity to work.

Example

Wide goals can generate spin-offs in other areas

It is said of development projects that "the more precise the target definition, the more efficiently the target can be reached". This may be the case for many development projects. For innovation projects though, the equivalent saying could be: "the less precise the target definition, the more effectively a target can be reached". With many innovation projects, a target in terms of product or process may not exist at the onset. Instead it is the aim of the project to identify a suitable product or process, as shown in the following example.

Bernie: How is the work on the low-cost generator going, Bill?

Bill: Fancy a look at my latest brainwave, Bernie?

Bernie: Sure.

Bill: Have a look at this printed circuit. Notice anything unusual?

Bernie: Well yes. It looks like the leaf of a tropical plant.

Bill: As it happens, it is not a souvenir from my last holidays. It is a printed circuit that does not fail even if you spray it with salt water or dip it in a glass of water. Incidentally, I have already protected my idea by patent.

Bernie: I am glad you picked up the suggestion from our last chat. But do tell me, what does this 'rainforest printed circuit' have to do with your project?

Bill: Nothing yet. It has however, already revolutionised our submergible pumps.

The innovation cell had hit upon an idea that improved a product other than the one the innovation cell was trying to develop. It pursued the idea, and gave it to the development team responsible for the product. It then returned to the original task at hand.

Here are some of the unknown factors you may encounter and be afraid of, because you probably don't know how to deal with them:

- Nonlinearity
- Unpredictability
- "Chaos" / un-orderliness / "messiness"
- Self-organisation
- Organic organisation

Few business schools teach one how to understand, influence and be successful in environments that exhibit these characteristics. Most businesses don't dare even touch the subject. But despite this unwillingness, organisations that are built on these premises are already emerging. Now what makes these characteristics so foreign to us that we are even afraid of them?

Example

Succeed despite risk and face the biggest change

Few people like facing risks. Risks can all of a sudden turn into real threats for a company and its employees. It is only too understandable if people like to turn the other way when risks are looming up in front of them. A company however, cannot afford to ignore risks, or limit its actions because of fear. The following example illustrates the irrational nature of fear, which may stop an organisation from moving forward.

Clare: I would like to show you my latest results with the AF institute on Sapphire.

Bernie: Any signs of a breakthrough in welding and moulding?

Clare: It now looks as if we won't even have to weld anymore. We can do away with the process step. This will raise the chances of moulding the parts without any scrap.

Bernie: You have been doing research on the Sapphire technology now for almost a year. I think that is a great result. What does the business unit say? They must be pleased.

Clare: That is just what puzzles me most. When I started the research, they were happy that someone was looking into new technologies for products and processes. Now they almost don't seem to want to know.

Bernie: Your technical solution six months ago was quite complex, and I was beginning to get lost when you explained it to me. Now with this breakthrough the solution has become simple enough for me to follow. Have you tried to explain the new solutions to the business unit?

Clare: Yes, Bernie. They seem to understand the new process easily enough, as you say. But I think it is dawning on them, that if my work succeeds, they will have to completely rethink their way of working.

The threat of change posed by the research project caused such a stir, that the business unit discredited Clare's work wherever possible. For a long time the business unit succeeded in cutting the project budget with the argument that if ever research proved the new technology to be viable, then the current business was jeopardised. The business unit was not prepared to finance its own internal competition. As a result it now faces very fierce external competition.

Non-linearity

James Gleick is reputed to have said "Non-linearity can be likened to the act of playing a game that has the effect of changing its own rules". When this principle is applied to an organisation, we may get what we call "agile" organisation. More than likely we will experience a different outcome than the one we had hoped for. The behaviour of the organisation may become unpredictable at certain times. Unpredictability is one of the most undesirable attributes of the efficient organisation. After all, order and predictability are the cornerstones of any reputable organisation. If these are absent, chaos lurks around the corner. Of course chaos stands in most cases for decay and doom. But it is a necessary prerequisite for the emergence of truly disruptive innovation. Remember that Penicillin was discovered not by an orderly process, but because of "messy" Petri dishes left unattended for a night. This same principle holds true for the organisation of innovative work. When we allow the group we selected, to do the innovative work alone at the right time, not intervening when it seems to get "messy", we allow self-organisation to emerge. Self-organisation will lead to the most natural and successful organisational form for a given innovation project. No formal leader is involved; no formal plans have been developed upfront. What is required is trust in the process, although from the outside we may not understand what is going on. Most managers and leaders will probably experience disgust at this approach, but it is exactly the approach we are recommending to the innovation manager. There are plenty of things to be afraid of.

If we trust this process, we will encounter success. Boldly stepping forward into this unknown world opens us up to the laws and rules that lie behind its messy surface. These rules and laws are more beautiful than we can imagine. Everyone who has seen the beauty of fractals or their compelling symmetry knows the order which always and without exception emerges out of chaos.

All we need to do to experience this is to trust in the capabilities of living systems and despite our fears, to have the courage to step deliberately into this world.

What should you take on the journey?

When we start the journey towards disruptive innovation, we need to make sure we travel light. We have already overcome our fear of the unknown, and now we need to think about what to take on the trip. Because the pathway is so unfamiliar to us, every bit of progress, even when small, is worth recognition and frequent acknowledgement. The celebration of small successes will be as important as the mourning of failures, which we will undoubtedly also encounter. Often we will experience the transformation of failure into success, which will give us a reason to celebrate.

Although successes will come in many shapes and sizes, most as specific variations of a given project, some generic ones to be expected are listed below:

- the joy of collective clarity
- the satisfaction of being able to deal with the unexpected
- the gratification of facing and resolving conflict

The joy of collective clarity stems from the group's need to talk about its trip continuously, so that everyone understands where the journey will end. This is not to be underestimated. Only when we truly understand each other, can meaningful conversation emerge. But clarity is also required for alignment to common goals. Instead of behaving as in the following story, we might actually be able to understand each other.

The story of the blind men examining an elephant

Once upon a time, five blind men were asked to examine an elephant. The first was given the trunk, the second the feet, the third the tail, the fourth the skin on the body and the fifth the teeth. They were asked to examine carefully the part they were given, without knowing that it was an elephant they were examining. After a thorough and lengthy evaluation, they came together at the end of the day to exchange their findings, and to reach a conclusion about what it was they might have been examining.

The first, who had examined the trunk, declared with confidence that it was nothing but a big snake. "No, no", said the second, "you are completely off track. What we were examining was clearly a group of trees, planted firmly in the earth". "Ha-ha", said the third, "what stupid men you are, because you are not seeing" (another joke maybe) "the obvious: what we had in our hands was clearly a poisonous cobra snake. I still shiver at the thought of her forceful movements". "I cannot believe the ignorance around me", said the fourth. "Has none of you experienced the roughness of the surface, the little bumps and hairy things sticking out? For sure this was a poorly-sewn tablecloth". "Oh Shiva", declared the fifth, "rid me of these blind" (another joke) "and stupid people. What a false statement about the roughness of the surface. Haven't you experienced how smooth and beautiful this thing really is? It can only be the masterpiece of a craftsman who created one of the most beautiful pieces of jewellery you can find". And so they engaged in an endless dispute, no one ever giving up his point of view, and never, ever experiencing the elephant for what it was.

This story may have reminded you of one of your last "all-hands" or management meetings, but for sure it is not what innovation teams will experience. This is because team members will have developed the ability to see clearly, individually as well as collectively.

Of course they will also face failure. It's not so important what specific kinds of failure they experience. More significant is the way they deal with failure. Whatever setbacks they experience, the team will always get back on its feet and learn from it. Therein lies its strength – the relentless drive forward, whatever it takes.

Your first step as an Innovation Manager

What makes a team take such a stance? Your role as a leader makes all the difference. Every journey begins with a first step, and you will be the one to take it. This first step is always unique to you. Your team will see this uniqueness and appreciate it. From then on you may have the followers you always dreamed of. But beware that your followers may become leaders themselves, during the course of your innovation project. Your leadership style needs to adapt to the maturity level of your team, and this will include once in a while, the need for you to follow, while somebody else leads. Here are some of the different possibilities you might encounter on your trip:

Sometimes the team will look to you for straightforward guidance. In this situation you need to tell them clearly what their options are, and which one, from your perspective, is the most likely to succeed.

In other situations the team needs to consult with you. Here your leadership becomes that of the advisor and guide. Your view will be listened to and considered, but in the end the team decides for itself what to do.

Probably the most difficult situation for traditionally educated leaders and managers is when the team simply asks you to follow it, without any "ifs" or "buts". Here you need to demonstrate trust in the team. It is a most demanding, but also very rewarding situation when handled properly.

Each of the situations includes a few more finely grained variations. But in essence, leadership on the way towards disruptive innovation breaks away from the traditional arrangement, where leaders and their followers are separated. It becomes dependent on context, where specific situations bring forth the leaders appropriate to them. This is hard work for you, because in the greater environment of the established organisation, this leadership approach will be met at best, with confusion. Because of this, something else is required of you. You will need to keep your traditional styles of leadership polished up, for communicating with peers and leaders outside the realm of the innovation project. This is a tough task, which will demand much of you. In the process you may even be "used up", much like a reactant in a chemical reactions.

Why then, should you engage in such an endeavour at all? There are the obvious reasons of creating the desired result and reaching the set goal. When the journey is over, you and the team can leave it behind and look forward to a new adventure. But this is not all. Simply the fact of having travelled the path of an innovation cell towards disruptive innovation will have a lasting effect on you. The trip itself may be worth more than the results it created. This is because while taking the plunge you will experience your original freedom and creativity as a human being. The more often you plunge, the better you will become at it. This is a worthwhile reward in the business world, where somebody else dictates most of what we do. So in some way, engaging in this journey is also a first step to becoming again an independent human being. This may be true for all of the people involved in taking the plunge.

□ *Well yes, I do want to innovate. I want to find fresh and unexpected territory. I want to go beyond existing boundaries to find those ideas that can make us successful in the future. I must admit, I am curious by nature!*

● Fine! But be prepared for a rough ride. At times you will be plagued with doubt, and at times joyfully elated.

□ *This sounds good, but scary! I think I'm about to enter the dark zone. Of course I'm feeling nervous. I will need help- can you provide me with guidelines?*

● Yes, there are guidelines. Oh, and by the way, in fact you are just leaving the dark zone and walking into the dawn!

5 Change is hard work

- When you start out to innovate, you are beginning a journey- and like any journey there are preparations to be made.

- *You mean pack a suitcase, find tickets and passport, make the sand-wiches and check the tyre pressure?*

- Yes that is a very good analogy. Think about Columbus. He acquired a ship, put together a crew and accumulated provisions well in advance of setting sail for India.

- *Isn't Columbus a bad example? It was pot luck that he found America, and it had taken him years to obtain the Duke's permission to set sail.*

- Well yes. But Columbus had prepared himself well. He was an expert cartographer. Obtaining the Duke's permission was a mighty act of di-plomacy. He was ready to deal with the unexpected. Even when the outcome was different from his expectations, he was happy to continue his explorations. This example illustrates the nature of innovation rather well. The timing has to be right, the appropriate authority has to agree, you must be prepared for the unexpected, you need to know your trade, have the courage and fortitude to continue in the face of obstacles, and you could do with a bit of luck on your side.

  Let's look at what this means.

Essay

Practical aspects of creating an Innovation Cell

An innovation cell is a specific form of organisation for an innovation project. It consists of a team of volunteers who are dedicated, have full control over the project and who will disband again, once the project is over.

Where do you set up an Innovation Cell?

Innovation cells are particularly suited to innovations and projects with high risks. They have proven themselves in the complex environment of the automotive industry. Innovation cells play a particular role in the development process, the organisation and the market.

Where in the development process do you best employ Innovation Cells?

You may employ innovation cells for the early stages of the innovation process, where the gap between idea and product is greatest. Here are a few typical situations:

- the technology is known, but the product is not
- existing products are produced with outdated technology, but it is not obvious which new technology might be suitable
- a customer is interested in a new product or feature, but the feasibility is not yet proven
- neither the technology nor the product is known, but there is a growing feeling that the product is required, and a reasonable hope that the technology can be mastered.

The innovation cell can start off working with the bare idea, and continue until the product is designed, or the production is set up.

Where in the organisation do you best set up Innovation Cells?

You may use innovation cells to create strong teams across existing divisions or departments. Here are a few typical situations:

- you can establish a new technology by pooling the skills of two different departments

- a new technology comes into consideration for one of your products, but it is only available outside your company
- you don't have the resources for a project, and need to pull in help from outside your department or company
- the team may continue to work together after the innovation cell itself has stopped, but the project continues

Where in the market do you best position Innovation Cells?

With innovation cells you can turn previously unattainable market opportunities to your advantage. Here are a few typical situations:

- you need a lead into a new technology or a new product
- your competitor is about to come out with a new product, and you need to follow quickly
- a new market opens up to you, but you can't afford to divert critical resources from existing markets
- you feel there is a new market opportunity, but you don't quite know which product would allow you to grasp it.

What is special about the Innovation Cell?

The innovation cell is a special but unusual form of project management. The success of the innovation cell depends on the following five properties, and their effects on project management:

1 The autonomy of the innovation cell

The team is in charge of its own decisions. Even the decision to abandon the project is allowed.

Example

In the case of reaching a dead end, the Innovation Cell has the right to stop

One does not have to resort to military examples to illustrate the tragedy that follows a futile mission forced to continue. For innovation projects, the following wisdom seems to hold: don't push it if all participants predict imminent failure. When the members of an innovation cell unanimously declare that the mission has failed, then believe them. The cell should be allowed to stop and disband. In the following example, management makes a mistake, and forces the innovation cell to continue without the necessary provisions:

Gerald: The work in the EcoLie project has made good progress and has now reached an important gate. We know about the technology, have established a network of suppliers and set up a plan for implementation.

David: I am glad to hear it.

Gerald: You might like to look at the implementation plan in detail. The innovation cell now needs your approval and the approval of the board for the next phase.

David: What is it we need to approve?

Gerald: We knew from the beginning of the project that the implementation would require heavy investment for the machinery, and also the enrolment of production staff.

David: You know full well that we can't spare anyone from production just now. Neither can I ask the board for further investment when only last week I asked for the extension of the prefabrication plant.

Gerald: Is that a 'No' to the innovation cell?

David: Not quite. You know that EcoLie is a prestigious project, which we need in order to keep our shareholders happy. There is no way we can stop it.

The innovation cell was denied the necessary working material. At the same time it was not allowed to stop. In other words, it had to continue and pretend to make progress. After nine months the agile members of the innovation cell had found jobs outside their own company.

2 The dedication of its members

Each team member concentrates his or her full attention on the project. No other job gets in the way. No other obligation takes priority. This demand sounds simple and easily satisfied. It is surprising, however to note just how difficult it is. In companies today, we find that every member of staff, with the possible exception of cleaners and apprentices, has to multiplex tasks. It is very rare for a standard company to allow an employee to pursue just one task, and it has become an uncomfortable and even unacceptable state for an employee to work only on one thing. The value not only of managers seems to depend on the number of assignments that are carried out simultaneously.

Example

A place for unexpected challenges

Coordination and synchronisation of many individually developed components into one coherent whole, is a significant challenge in many large software development projects. Technology assists in the task, but successful component integration is equally dependent on the discipline of each individual developer. Sometimes, developers have a "last minute" important idea that they think simply must be a part of the next integration cycle. The problem arises when these "ideas" are introduced into the code at a point in time when the integration cycle has already started, and the new piece disrupts the whole synchronisation process. As a result, the work of the whole team might be jeopardised. In many teams this challenge is not addressed openly, because the developers who demonstrate this behaviour are often also highly respected for their technical skills.

DevTeam experienced the same challenges, and resolved them in a rather unusual way. The following conversation during an "all hands" meeting between Martin, the project manager, and Lee, one of the technical lead developers, demonstrates how a simple idea helped in this critical task, although it was not specifically targeted towards this problem:

Martin: I would like to introduce a simple tool allowing us to express our feelings when something that happens in the team or around us distracts or angers us.

Lee: Wow, now you really are getting crazy. This is a high-technology project; there are no emotions involved. We are all professionals and trained to focus just on the rational aspects of work.

Martin: Yes, this is what you and the other team members tell me all the time. But do you know how often one of you takes me aside to tell me just how idiotic other team members, or the people around us- specifically the managers, are being? I'm tired of all this, and I think there is a better way to work out these situations. Do you want to hear about it?

Lee: OK, what is it?

Martin: We will hang this flipchart paper next to the status graph in our team room. As you can see, it is empty except for the title. I call it the "Unexpected Challenges" chart.

Lee: What do you expect us to do?

Martin: It is simple. Whenever something happens which upsets one of you, you write it on this flip chart paper. You can do so anonymously and whenever you like. Our team room is never locked, so you have access to the chart without any restrictions.

Lee (laughing): Martin, you must be on dope! This thing will stay blank forever. But it won't do any harm I suppose. Let's hang the chart up.

Within a week the first chart was full of comments, among them the following:

- having to use XY Project, and having to report every minute detail to the directors
- interface credibility! We are too optimistic, as usual
- system update!!!

Behind these comments, some of them anonymous, others signed, stood a whole story. The "System Update!!!" comment came from Lee. One of his project responsibilities was code synchronisation. He did this task at night, to prevent interference with ongoing development work during daytime. During synchronisation no one was supposed to send individual code to the integration system, because this interfered with the process. Lee would send an email around alerting everyone to the imminent synchronisation. But sometimes team members sent their code anyway, often assuming he had not yet started the process. In the incident that triggered Lee to write the comment, he had lost a whole week of work. One of the developers had made a tiny last-minute change in an interface, which affected 90% of the whole system. He was truly upset.

His comment made the problem visible to the whole team. Because of this, the team as a whole talked for the first time about possible ways to prevent this situation. The solution the team agreed upon was simple and effective. Instead of sending an email, which people often did not read in time, Lee would raise a "Jolly Roger" flag above the developers' cubicles at a place where everyone could see it. As long as the flag was up, sending individual code was forbidden. This approach worked right away. Interference with synchronisation activities was never a problem again.

3 The co-location of the team

The team works in one room. The team members originate from different units. They will typically arrive bringing different knowledge, experience, responsibilities and social skills. When they leave the innovation cell, everyone will have learned something from the others.

4 The goal orientation of the Innovation Cell

The common goal holds the team together. The team is responsible within the set conditions for the project. The team learns to live with the consequences of its own decisions.

5 The Innovation Cell is ephemeral – it's over when it's over

The team only exists for the duration of the innovation cell. At the end the innovation cell is disbanded, and all members return to their original units.

What skills are needed, and what will you receive in return?

The innovation cell works on the principle of trusting in competence i.e. you as the manager trust the team, and the team tries to live up to your expectations. You as the manager have to give the innovation cell its sense of purpose. The innovation cell derives its sense of purpose from the will to succeed. The innovation cell raises the level of achievement each individual team member wants to attain. The group dynamics help each member in the process of orientation. Very soon they will either find that they are enthusiastic about the common goal, or they will not be accepted by the group, and therefore will want to leave.

You will have to ensure that the innovation cell has the authority to act. The innovation cell increases the ability of all the members in the team. Close proximity and the pace of progress create a learning atmosphere. The team members will grow into their respective jobs.

You as a manager have to grant full decision rights. The innovation cell will then make decisions easily and competently, as well as reacting quickly to external influences.

The innovation cell is relatively free of superimposed rules. Within the confines of the project goal it is able to organise its own environment. Its closed nature enables it to experiment with new ideas, to try things that haven't been tried before and to create something new. In the innovation cell you are permitted to make mistakes, as long as you are prepared to learn from them.

What are typical results achieved by Innovation Cells?

The innovation cell is given a challenging goal. It delivers results, which may be unusual. In particular, the results may be disruptive i.e. bring out new products and technologies that may be considered threatening by others. As a manager you should be aware of the possible emotional side effects, and not let them deter you.

1 Quick development

Due to its autonomy and full dedication, the innovation cell tends to outperform the standard project management team. Because of close communication the co-located team is not only more creative, but also faster in the acquisition of knowledge and skills.

Example

An ambitious goal mobilises energies in a closely-knit team

The saying 'When the going gets tough, the tough get going' seems to hold for innovation cells. There is nothing like an ambitious goal or stiff competition to mobilise the energies of a closely-knit team. This came true for the LinearDrive innovation cell in MyCorp, when all of a sudden the customer TransStar gave them the opportunity to prove that they could beat Other-Corp, a competitor.

Hugo: Last night I just got back from TransStar. They let me know how disappointed they are with one of our competitors. From the way they talked about it, it had to be OtherCorp.

Bernie: It is nice to know that our competition gives TransStar cause for concern.

Hugo: Well, that is not all. TransStar is asking us whether we can deliver LinearDrive better than OtherCorp. I reassured them that we could, even though I remember that we set our own project on hold.

Bernie: Then we now have a unique opportunity! You are right; the project was suspended only six months ago. But I am sure we can start it up again. Will you help me jump-start an innovation cell today?

Hugo: I would be delighted. Can I tell TransStar we will show them a demonstrator that will outperform the one from OtherCorp within the next three months?

Within four weeks the first prototypes were tested at TransStar. The complete innovation cell team was there. The same day they met the engineers from OtherCorp. The race was on. Three months later, the contract for LinearDrive went to MyCorp. The innovation cell finished the development, and was disbanded only after the successful launch of the product.

2 Products that access new technologies or markets

A typical challenge for innovative companies is how to enter a new technology or a new market. Usually the effort involved is greater than the reward from the first product. The innovation cell is able to deliver the first product (or 'enabler') quickly, and use the momentum to launch the second, and hopefully more profitable product, soon afterwards.

3 Unexpected benefits

Spin-offs are often the nicest surprises an innovation cell comes up with. The excess creativity paired with the fast growing knowledge in the innovation cell may lead to unexpected results. What was initially considered to be a mistake may turn out on closer inspection to be a valuable contribution. Spin-offs can show the hidden potential of new technologies or markets.

Example

The Innovation Cell provides a feeling of security

You may have a bright idea, but are you brave enough to mention it? You will be more likely to surrender an idea spontaneously, if you are very familiar with the people surrounding you. An innovation cell is like a family. The following example shows that the familiarity within an innovation cell relaxes the guard on spontaneous participation, and fosters a spirit of free interrelation. Hugh, Jim and Tom are members of the innovation cell. Chris is helping them with their demonstrators, but is obviously not happy with his job. Listen to how unusual ideas come up in the innovation cell:

Chris: I am getting fed up with this complicated piece of electronics. We need to tailor each time we integrate a prototype into a customer demonstrator.

Jim: Well, do you want to buy a general-purpose personal computer, adapt the software and hide it in the demonstrator each time?

Hugh: No that would be both too expensive and too bulky.

Tom: Why don't we use the electronics out of some toy?

Jim: That reminds me. My son had a new Lego set only last Christmas. It contains an electronic controller.

Thom: I think I know the one. We used them in school for our electronics class.

Jim: I 'll bring it round tomorrow.

Tom: I can't wait to try it out.

Chris: Well, I will be around tomorrow to see what you can conjure up
         with the building set. Just imagine what my boss will say, when I
         ask him to sign the purchase slip for a toy set.

The ease with which ideas are accepted, played with and rejected is
amazing to an outsider like Chris, who really came into the innovation cell
to vent his frustration. He didn't anticipate the crossfire of ideas. He leaves
the innovation cell with new hope, and the chance to surprise his boss with
the idea he just picked up.

4 Early knowledge of potential failure

If there was no result to be had, then the innovation cell will show you the
limits in the original goal definition. If the goal can be changed, the
chances of a positive result can be raised.

How does an Innovation Cell differ from a standard organisation?

The innovation cell differs from the typical unit within the organisation in
a number of areas:

- Process-orientation versus result-orientation: the innovation cell tends
  to be more result-oriented than a typical unit in the organisation

- person versus task: the innovation cell pays attention both to the people involved and the task at hand,

- organisational versus professional: the innovation cell is less affected
  by concerns about conforming

- closed or open: the innovation cell marries the close nature of the pro-
  tective cell with the open mindset for innovation

- strong control versus weak control: the innovation cell exerts less control on the team members than the traditional unit in the organisation

- normative versus pragmatic: the innovation cell tends to go for the
  pragmatic approach

The innovation cell complements the nature of the more traditional organisation.

Example

A weekly routine to foster self-responsibility

The following example shows how the innovation cell develops a habit out of a practical need.

Sixty developers have very little time to solve a very complex problem and create a product that after two unsuccessful development years is seriously behind schedule. Past attempts, such as traditional project management, have failed to bring the work back on track. The project team is demoralised and is looking to George, the new project manager to solve the problem. George is convinced that heroism and belief in the "superpowers" of just one person are an illusion and will not work. Based on the history of the project team, he knows that a lack of frequent and in particular honest feedback on the state of the project, contributed to its current desolate state. In one of the first team meetings this challenge is being addressed in the following way:

George: We are meeting today to talk about a process by which we know for certain what the state of the project is. You realise that not knowing the state of everybody's progress was a major cause for the delay of the product.

Frank: We used XY Project all the time but unfortunately I never received accurate feedback from the team members regarding their own progress. That's the reason why the project plan looked OK for a long time although we were already missing milestones.

George: What do you think contributed to this situation?

Randy: You know very well how we work here at BSWC: first we over-promise, then we hit a wall, and finally, we select a hero who will solve the problem and get the rewards. Everyone else gets punished. Making a mistake in this company is always dangerous.

Frank: I am really concerned that if we continue in this way, we will fail again. We need to find a process that keeps us honest regarding each individual's work progress, even if it hurts.

George: The weakest point in the previous feedback process appears to be that there was no separation between contributor and assessor of the contribution. For example: if Randy is responsible for a certain task and he is also the one who assesses his own work, when something goes wrong he will have an internal conflict about reporting the truth. Taking the BSWC culture into account makes this even worse.

Randy: I appreciate your candour, but how do we get out of this vicious circle?

George: I suggest we do three things:

We agree on a regular cycle of meetings where we meet and talk about project progress.

At the beginning of the cycle everyone states what he or she will have achieved by the end of the cycle.

At the end of each cycle the whole team assesses the state of each contribution, instead of leaving this to the contributor. This may be painful to start with, but it will help us to stay honest.

DevTeam decided to go ahead with the proposal and used a weekly cycle as its base project rhythm. Each Monday morning the whole team came together, organised into smaller groups representing the main constituencies of the project. Every team member stated what he or she would create during the following week. On Friday mornings the same groups came together, and in each group team members assessed each other's work. This feedback was put into the project status report and the project plan.

The initial meetings were painful. Few people had ever gone through such a rigid process. But after the third cycle the benefits became clear. Honesty, clarity and a sense of reality became the norm for the team. The breakthrough came when most team members realised that assessing their *contribution* was different from being personally assessed. The team created this separation by using specific phrases such as "In my view the contribution has this or that status because of the following observations…" or "My suggestions for improving the task are…" when in the assessment meeting. By the end of the successful completion of the project a methodology had emerged that proved useful in many other feedback meetings in different teams and projects. It had the following basic outline:

Each team member prepared individually in a preparation phase before the feedback session began. The following questions were used for guidance:

- What is the purpose of my feedback?
- What relevant facts and observations can I use in the actual feedback session?
- What are my reactions to the status of the work I assess?
- What suggestions can I make to the team member whose work I assess?
- What do I need to remember about myself when providing the feedback?

Team members made notes in a simple form for each of these questions. In the actual feedback session following the same questions, the notes were given to the person whose work was being assessed as an additional means of providing input.

How will you and your organisation benefit from the Innovation Cell?

The innovation cell can act as a catalyst for change. When you employ innovation cells in your company you are likely to:

*   create a new form of communication
*   increase the tolerance to failure
*   promote those willing to change
*   transform your organisation into an agile innovation organism
*   create parallel structures
*   raise the willingness of the management to change by removing monopolies on information, reducing absolute power of position and raising your chance for creating successful disruptive innovations

Most importantly you will create new success stories.

5.1 Prepare for the journey

*   The first thing to do is to identify and collect your resources.

☐ *What is that- people and money?*

*   Not only people and money, but also the right sort of people and money. You need people with the right skills and the right values.

Example

Personal stories as a means for team selection

Imagine you have been asked to form a self-organising innovation team. Your company's leadership team supports your task and departments are offering their help. What do you know about the people who are interested in working in your team?

There are processes that allow people to share their view on questions that matter in an authentic way. Listening to each person's story gives a first hint as to whether, and exactly where they will fit in. This is an unusual process, requiring that the leader be fully behind it. The following dialogue between Charles, the leader of a newly forming innovation team, and Carl his external consultant, is representative for the initial phase:

Charles: There will be many people who want to work in this innovation team. What would help me in my selection process?

Carl: There is a process that enables people to share their ideas and beliefs openly without too much effort. It is basically a story-telling process based on certain questions. Listening to these personal stories will give you a first indication of who may fit into your team.

Charles: "Story-telling" sounds esoteric. How does it work?

Carl: Imagine a piece of flip chart paper on which is drawn a circle divided into 4 quadrants, and below it an empty text banner. Each of the 4 quadrants has a question assigned to it that the participants have to answer individually.

Charles: So? What's so special about it?

Carl: Well, for one the questions are pretty personal. In your case you could ask questions such as:

1. What are you proud of as an innovator?

2. What is your contribution when working with others?

3. What hinders you the most in your creative work?

4. What is your dream for innovation teams?

Charles: Aren't people going to write lengthy essays when answering the question?

Carl: Here is the trick. People are not allowed to write any words in the quadrants. They have to draw!

Charles: You mean like children? You must be crazy! Adults would never go along with that!

Carl: Rest assured. This has been done worldwide many thousands of times. I myself have done it with hundreds of people – it has never failed.

Charles: Well, I still can't visualise it, but we have worked together before - on this one I may just have to trust you. But one other thing, what about the text banner?

Carl: Oh yes. This is the place to write a short sentence summarising the basic idea behind the pictures the participants have drawn.

Charles: OK, let's assume people have done their pictures. What then?

Carl: People will share their pictures with you and the other participants in a group setting.

Charles: And how do they do this?

Carl: They pick up their picture and post it somewhere where everyone can see it. Then they explain the picture to the group in their own words. This is the story-telling part.

Charles: You have done this many times. What is the result of these sessions?

Carl: People see it as an opportunity to share what is really important to them. The request for drawing the answers makes them think more deeply. Often it is the first time that people have a chance to reflect consciously on these topics. It always results in individual gratitude for the opportunity and very high group cohesion. People basically see themselves with new eyes – and they appreciate it.

Charles: OK, let's do it.

The actual activity takes about 1 - 2 hours depending on the group sizes. People always report that after this session, the conversations in their teams have a more open and deeper tone than anything experienced before in similar business settings.

When using this process for selecting people for a certain task, it allows both the leader who is looking for someone, and the professionals who are interested in the project, to see more clearly if and how they fit in.

- You also need people who are interested and even very enthusiastic about what you want to explore. Thus the heart and mind play an equally important role.

☐ *Do you mean hard and soft skills?*

- Yes. For the team to function properly, the soul of the group needs to be addressed. This poses a question: can the team develop a collective spirit, or will it remain a cluster of individuals?

☐ *Does this mean I can only gather people into the team who are comfortable with each other?*

- It depends what you mean by "comfortable". If your understanding of "comfortable" is "harmonious at all times" then the answer is no. Rather the opposite. For a team to work effectively as one, it should be made up of people who differ considerably from one another.

□ *How do I glue them together? I can see them splitting apart as soon as the first conflict arises in the innovation project.*

- This is where you will need leadership - and I don't mean management. When a conflict arises, your contribution as a leader is to provide an environment where the team can resolve differences without splitting apart. Are you a leader? Do you want to become a leader?

5.2 Focus your effort

□ *How can I become a leader?*

Essay

Can I manage change and stability side by side?

By now you might be ready to contemplate launching into a disruptive innovation. You may have a project in mind and visualise a team of experts working away at it. But what will happen when they encounter their first stumbling block? And more to the point, what might your colleagues say, when they notice that you are using unconventional methods? You will have to prepare your move and consider carefully how to embed your innovation project in the company, its structure and tradition. The innovation cell can help you.

How to bridge the natural incompatibility of the hierarchical organisation using self-organisation

First of all, you as a manager will have to care for the team members in the innovation cell. If they are not familiar with the innovation cell concept, you will have to see to it that they become enthusiastic, or at least curious about the goal and the method. Make sure they know that any results they produce will be theirs. Show them that you are ready to support them, and that you will stand behind them with your authority. They will want to know that they can concentrate fully on the job in hand. You can help them to clear their paths, which otherwise might be strewn with small bureaucratic stumbling stones. Experience with innovation cells shows that once the formalities recede that rule so much of the daily life of standards development, then the team is ready to understand and incorporate a challenging goal.

Experience shows that innovation cells need a few incentives. The most important incentives are the freedom from the tangled web of bureaucracy described above. Then comes your attention. The members of the innovation cell need to feel that the project is important. Go and see them. Ask your boss to pay the team a visit. For many team members, top management attention is reward enough in itself.

You must also ensure that your colleagues support you and the innovation cell. The ephemeral nature i.e. the fact that it will come to an end, will help you. It is easy to convince your colleagues to part with a valuable expert, if you promise to return the expert after say, nine months. It will also help if you inform them that once the major risks are removed, the project will run in the conventional manner. In other words the innovation cell will not threaten the standard development and production process of the company, but only augment it, pointing towards more unusual and challenging goals.

How to deal with the unknown in a deterministic environment

In a company with rules, standards and regulations it will feel strange to be in a project with many unknowns. The sudden freedom may feel uncomfortable. As a manager in charge of innovation, you may choose to use the innovation cell to create new habits that allow you and others to confront the unknown. You can start by marking special events such as the project kick-off (and later the end of the project). Why not celebrate when the innovation cell has been successful. Why not have a 'funeral party' when the innovation project has to be abandoned. You might use these occasions to create a 'club of pioneers' i.e. a group of people who enjoy venturing into unknown territory.

On occasion, you may have the opportunity to raise the likelihood that unknown things will happen. For example, you can mix team members from different backgrounds, skills and even character types. Choose members from different company locations, different cultures or even different companies. Include an engineer from a key supplier or a student from a university. Experience with innovation cells shows that it helps to have a young and inexperienced team member next to an old and experienced one. Balance the team with respect to creativity and learning.

Example

Group dynamics at work in an Innovation Cell

A room full of similar people may be socially satisfying, but only when there is a good mix of different characters and abilities, does the creative nature of group dynamics get to work. For innovation cells this mix is decisive for the success or failure of the innovation project.

Hugh: How did you get on with the simulation of yesterday's model?

Tom: I will have the results back from the institute this afternoon. Yesterday my friend there gave me some information on a ceramic substrate, and asked whether we had considered it as a substitute for the metal layer.

Jim: Oh no, not ceramics again! We tried ceramics in the last project and it turned out to be far too expensive. Nearly sunk our project.

Hugh: Hang on. Take it easy, Jim. Give Tom a chance to explain what he means. We could do with a substitute for the metal layer. If ceramics is a possibility, I think I might just know the company to help us.

Tom: Let me try to explain what my friend suggested. ...

The innovation cell is held together by Hugh, who is calm, collected and seems to have an established network. Tom is a young student, hoping to prove himself in the innovation cell. Jim is the critical engineer, who needs to be convinced in order to accept a new idea. The mix of 'soft skills' helps the team sort out any relationship problems they may encounter. Together with a good mix of 'hard skills' i.e. technical competence, they stand a good chance of mastering innovation.

Example

Making cultural differences work for you

The intimacy within an innovation cell does not have to be there right from the start. In fact, it is sometimes better if it arises in the course of the innovation project. When the team members are new to one another, there is a chance that everyone will create his or her own role in the team. The following example shows how cultural diversity provides both challenges and opportunities. Brian has been in the company for a long time. Rangi is a foreign student, doing an internship in industry.

Brian: I noticed you are using tools from a public domain.

Rangi: Sure. These tools helped me a lot when I was doing my university degree.

Brian: No one here uses them. Are you sure that all that public domain stuff works correctly?

Rangi: A friend of mine supervises the domain at the university. His network comprises some of the best-known tool-suppliers.

Brian: To be "the best-known" is no guarantee though, is it?

Rangi: It isn't. But wanting to remain "the best known" is. There is a code of honour in the network, which I aspire to myself. But why don't you just try it yourself.

Brian: I just might. Can you help me getting connected?

In the example, Brian opens up to Rangi, and is all of a sudden ready to accept the advice of a junior. This in turn raised Rangi's credibility with other senior members of the team. In the end Rangi was leading the tool change in the innovation cell. By the time Rangi left, the other members had completely assimilated Rangi's way of working.

There are many ways to make handling of the unknown a comfortable experience. You may choose to use techniques such as "open space" for a strategy workshop, or "brain writing" for idea generation, just to name two.

How do you resolve power issues of manager and team autonomy?

There are many books written about the resistance to change in companies. Many of the effects described also occur in connection with innovation cells. There is however, something special about innovation cells and the disruptive nature of the innovations they enable. The innovation cell acts within the company i.e. the threat of change usually perceived to come from outside, comes from within the company. The innovation cell may cause a conflict not only of interests but also of power amongst the management.

The conflict is easy to resolve if you learn to distinguish between two types of power. It is common practice amongst managers to define their power by the number of people working for them. In other words, management power is power over people. The innovation cell is a team that creates and shapes a new product or technology i.e. the innovation cell has power over creation. In order to be successful, the innovation cell needs to pick up this power over creation. At the end it will need to let go of it again.

The power over creation is what the innovation cell has in common with the entrepreneur. Power over people is what management has in common typically with a military organisation. Management may therefore look at the innovation cell with envy, but not necessarily with fear. As a leader you can distinguish between envy and fear. Use the distinction to create an innovative culture.

How to encourage failure tolerance in a lean company

As you embark on the road to innovation and change in the company, you will be leaving the well-paved roads of what is considered right and wrong. You will encourage the team to go where no one has been before. In other words you will be making mistakes, some of which will turn out alright in the end, most of which, however, will remain with you in your learning experience.

At first the isolation of the innovation cell will help you. With no one able to watch closely, you can save yourself a lot of unnecessary criticism and allow yourself and the team to get on with the innovation project. There are many methods, such as "The Five Why's", which can help you and the team to learn from mistakes. Dig down and find out why things are not going as you were hoping, and you will arrive at a better understanding of the situation. Check whether there is something to be gained from your mistakes. You may not always invent the "post-it note", but you might nevertheless use the knowledge and experience gained for future work.

The lean organisation will have procedures set up which will control and monitor most moves. They are unlikely to provide the freedom necessary for experiments. Use the innovation cell to encapsulate and guard the freedom required for creation. Don't check every small step that might make the team accountable for the overall goal and the budget. Later on the rest of the organisation will realise that the innovation cell takes on the pioneering role for a company of settlers.

How to accommodate the real possibility of personal failure

Don't expect everyone to come running to your innovation cell. The innovation cell will have its attractions: a new challenge, new colleagues and an otherwise unknown freedom. But it also has its drawbacks. The potential member may ask what he or she will receive in exchange for leaving the safety of a known environment. You may have to offer incentives that will attract the sort of people you are looking for. You will have to ensure that regardless of whether the innovation cell is considered ultimately to have been successful or not, its members do not lose face to the rest of the company.

What incentive can you grant members of an innovation cell? The answer is: few, and only the right ones. In an organisation geared towards efficiency, incentives play a major role. A typical lean organisation will reward the most efficient workers with a bonus payment. Managers can be encouraged to make the organisation even more efficient using the promise of career advancement. The innovation cell, however, is an organisation geared towards effectiveness. Whereas changes in efficiency are easily measured, changes in effectiveness are not. In addition, having more money or power will be of limited attraction to a pioneer. A pioneer is more driven by curiosity, idealism and holding the attention of those in power. So you should choose your incentives accordingly. For example, make the innovation cell a place where top management feels welcome. The innovation cell will reward you with performance, and its members will know that their effort is recognised. This will help to get over a sense of failure, should the team realise its first mistake was to reverse out of the cul-de-sac of an oh-so-promising development path.

Example

Reward and praise has to suit the individual

Progress in the innovation cell comes from all members. Whereas some members will push to the front to receive the credit, others prefer to remain in the background. Bernie, a manager for R&D and driver for innovation in MyCorp, strives to readdress the balance, and ensures that praise reaches all members.

Bernie: This autumn we are planning a technology workshop at our subsidiary in China.

Michael: I always wanted to go to China.

Bernie: I thought you might. We could do with your help in the workshop.

Michael: Trouble is, my English is pretty rusty.

> Bernie: I am sure we can work things out for a translation. In any case, I'd rather have the authentic expert, Michael, than just any old fluent speaker.

> Michael: I can contribute the latest developments we made in the innovation cell. Also, one of the students will probably help with the English. Do I need a passport?

> The trip to China was a great treat for Michael. He will probably be telling his grandchildren about his great trip. He certainly became the celebrity of the workshop, as his authenticity, knowledge and humility struck everyone.

The innovation cell is a place for individual learning and even the gaining of new qualifications. For example, it is possible to make participation in an innovation cell part of a company training scheme. Or you can hand out trophies and certificates, which commemorate and value innovative project work.

How do I learn a sense of timing?

It is your job as a manager to determine the right time to launch an innovation cell. For the innovation cell there is hardly anything worse than a false start. The innovation cell can be likened to a boat. Before launching a boat you have too ready it, in other words you have to fill it with provisions, make checks on all the machinery and work out the best time to embark. You should proceed similarly with the innovation cell.

The conditions are right for the launch of an innovation cell when the market is almost ready for a product or technology, and if the innovative task is really worthwhile. Please note that all the other rules for product timing hold, such as not cannibalising your "cash cow".

✓ The product has a chance in the market,

✓ the product is technically feasible, even if not all the remaining risks have been resolved

✓ the product enables new synergies within the company

✓ the task on hand is sufficiently complex to demand special attention

✓ bringing together the skills in the team will create new skills for the company.

The innovation cell is ready when all the provisions have been made, as follows:

✓ Make sure everyone who is important to the innovation project knows about it, and see to it that you have the support of the decision-makers in the company

✓ define the goal for the innovation project using external references where possible, such as customers, suppliers or law-making bodies

✓ put the team together, and pay particular attention to the mixture of competence and character type

✓ identify the support team

and now you can set off!

When should you stop the innovation project and look for a safe harbour? Recheck the conditions for starting. If one of them no longer holds, you should either move on to run the project in the standard organisation, or abandon the project entirely:

- If the product has no chance in the market your most likely course of action is to abandon the project

- if the product appears no longer technically feasible or the remaining risks can not be resolved, then abandon the project

- if the product is technically feasible and no risks remain, then use the standard organisation and disband the now redundant innovation cell

- if the product brings no new synergies into the company, you may choose to stop the project or continue it with a partner, with whom the synergies can be utilised

- if the task in hand is no longer particularly complex and does not demand special attention, then use the standard organisation and disband the innovation cell

- if it is not possible to bring together the skills in the team or to create new competencies for the company, then consider your effort wasted and "go fishing" instead

Note that any of these adverse conditions may rock the boat sufficiently to force the innovation cell to grind to an early halt. As the manager in charge, you need to examine the situation, probably observe it for a little while, and use your common sense for the decision of whether to continue or not.

- There are three basic requirements for becoming a leader:

   1.   Respect the values of your team
   2.   Keep your team together
   3.   If the team say something can't be done, then listen to them

   A leader must remember- whatever happens, always respect the individual; respect the team and its values; when your team needs you, go out of your way to help; always listen to the team, even when they are saying things you don't want to hear; understand that you too are part of the team, and that you have a special role in maintaining the sort of environment in which the team can perform its tasks best.

☐ *This sounds like a very abstract laundry list!*

- OK, here is the practical version: gather your team of volunteers, put them in one room with one goal, and visit them at frequent intervals to see how they are getting on. We call this setting up an innovation cell.

Essay

Why one room?

It may not be immediately obvious why an innovation cell needs to be in one room and one room only. You may ask yourself the question: could the team not be spread over a number of locations, where each member resides at the location best suited to his or her need? Modern tools of communication would make it possible for the team to work together. If the task set is continuous innovation, then the team is likely to succeed. For disruptive innovation, however, the team should reside in one location only. Here are the reasons: the proximity of all participants, the joint orientation and the time constraint generates an atmosphere that encourages cooperation and progress. Here is how it works:

Group dynamics

The full dynamics of the group can only develop in close, physical proximity. Take for example tremendous enthusiasm. I may be able to talk about my enthusiasm to my colleague when I'm speaking on the phone. But when my colleague is in the same room with me, he is likely to become aware of the many facets of my enthusiasm. My body language will reveal all he needs to know about my comfort levels in a particular situation. For every new value judgement I make, my body will communicate the corresponding confidence. Therefore the project team, sitting in close proximity, can observe one another. They will have a better awareness of the opportunities and risks of the project than a "group" which is spread over a number of separate locations.

Close proximity will also foster a sense of mutual understanding. When spread over several locations it is difficult to adjust expectations realistically. Inadvertently, I might overwhelm my team colleague in a distant location with a certain request. But if he is sitting right next to me, I will soon find out whether my request is reasonable or not. Thus the team sitting in one room is more likely to balance the workload than a team spread over separate locations.

If the team is spread over several locations it is easy for one team member to remain an individual and not actually join the group. When gathered in one location, the team can rely on the contribution of every member. This is particularly helpful when different viewpoints are required to assess a new situation. By sharing viewpoints, the team will acquire a better picture than any individual could provide on his own. The team can thus grasp opportunities that lie outside the reach of its individual members. The team provides the opportunity for "reframing".

Learning

Being in one room together, each team member has the opportunity to learn from any of the others. First of all learning may be a simple matter of copying. Learning is a very personal process. Here is what may happen to the individual:

As I see what my colleague is doing, I may remember his actions, memorise contacts, or copy behaviour. I have the opportunity to ask for explanations and use his knowledge and talent to expand my horizons.

When in one room, I will notice the behaviour of my team members. I will become aware of their strengths as well as their weaknesses. I have the chance to learn to understand them better; I may walk in the shoes of the other person.

When in one room I am also unable to avoid uncomfortable work. For example, if there is one part of my work which I loathe doing, I may try to delay it as much as possible. The group around me, however, will stop me from being selfish. If I don't want to attract the wroth of the group, I will have to buckle-down and do the work. Another example is the acceptance of uncomfortable facts, particularly about myself. The proximity of the group will let me know about any annoying little habit I may have, or imperfection I don't want to face up to.

Working together as a team in one room will be a learning experience for every team member.

Address

If the project team works in one room, it will present one address to everyone outside. Even in an age when the Internet provides a platform for communication, to have one physical address is still an advantage. First of all, with more than one member in the room, there is likely to be a contact available for outside calls. For customers or suppliers the team will be available, even if only one member is in the room. This availability has two effects, one for the team and one for the outside correspondent. For the team, availability implies that all team members should have the same information at hand i.e. communication and learning is fostered. For the outside correspondent, availability will further trust in the team and its project.

Example

The Innovation Cell assimilates its surroundings

In many companies the location of your office reflects your status. If for example, your office is close to a centre of power, then others intuitively respect your influence. For the innovation cell, the location is just as important. In the following example the location within James's testing laboratory was chosen deliberately for an innovation cell, to raise the expectation of 'quality innovation'.

James: I am concerned about the effort involved in setting up the new AllBright innovation cell in the testing laboratory. Do I really have to move people out of the testing laboratory and sacrifice much treasured lab space? Can't we leave the engineers from development and research in their office area?

Bernie: You probably know, James, that you are not the only one who finds the location of the innovation cell inconvenient. The engineers would also prefer to remain in their office area, preferably even at their separate desks in separate buildings. You know what it is like when engineers, who should work together closely, actually work in different buildings. I should think you agree with me, that we should move the team into one room together.

James: You must be referring to the last project for the MC 2000. The development was an absolute disaster. The first time the engineers met was when they were fighting in my office over whose fault it could have been. No, I quite agree with you, Bernie. The engineers should definitely move together for the sensitive concept and design phases. All I am asking is not to move my laboratory around.

Bernie: The difference between the MC 2000 project and AllBright is, that in the case of AllBright we are developing a new technology and a new product at the same time. This means all the quality criteria and test procedures will be new to us. Knowing the competence and dedication in your testing laboratory, I am confident that with the innovation cell in the middle of it, any new technology coming out will be well qualified and respected by internal and external customers.

The personal involvement of the head of the testing laboratory led to a product with a new technology, which was tried and tested long before it reached development and production. More importantly, it was accepted by development and production without any of the prejudices usually accompanying results coming straight from research.

For everyone in the company who is only indirectly involved in the innovation project, the address is a means to easy access. For example, if top management wants to know the state of the innovation project, then all they need to do is to pay a visit. The fact that the team room is open for inspection has two effects, one for the team and one for people outside. For the team, the surprise visit, particularly from the top management, is an incentive. For any outside visitor, the openness of the team room is a sign of confidence.

Selective access

Working in one room does not mean that the door has to be open all the time. In fact, it may be useful to close the door occasionally. It needs to be clear that being in the room means being a part of the team. The team room is not an area where anyone can walk in and out as they please. Only full-time team members have 'ownership' of the room. They may choose to keep some people out. For example, the innovation project may be kept secret, and not be general knowledge for everyone in the company. Or the project may be in an early stage, too early for uninvited onlookers. The project may be going through a sensitive phase, where uninvited guests could hinder it. By being in one room, the innovation team can decide who to let in and who to keep out. Thus the room provides a "safe space" for the team.

Spontaneous communication

Spontaneous communication occurs only within a small radius. When two people work in one room, spontaneous communication is the norm. It is easier to exchange information as and when required, rather than convening a meeting. When two people work in the same department, but not in the same room, spontaneous communication may still occur, but it is much rarer. If the two offices are over 100 metres apart, the chance meetings peter-out completely.

Example

Spontaneous communication leads to creativity

Communication modes vary with project tasks. When we are looking at the conception of a new product, spontaneous communication is particularly important, as it provides vital input for creativity. The chances for spontaneous, rather than scheduled communication depend on short distances or familiarity within a network of people. In an innovation cell the short distance is achieved simply by the location of people and meetings within one room. The mix of team members from different departments or even companies, as well as the mix of experience in the team structure itself, create the familiarity within networks of people, as the following example illustrates:

Tom (on the phone):.. if we use a solenoid instead of a dual position magnet, then we..

Jason (overhearing what Tom says), asks Sam: Didn't you talk to purchasing yesterday about the price for solenoids?

Sam: Sure, solenoids are one a penny.

Jason: If we over-dimension the solenoid, we can do away with a sizeable chunk of electronics.

Tom (covering the mouthpiece with his hands) to Sam and Jason: I think I am on to an interesting technical alternative here.

Jason: I think we have picked up the tune already. Could you ask your colleague at the university to run an analysis for the solenoid as well?

People within 10 metres of each other not separated by a wall, will communicate spontaneously at least once a day. The chances of spontaneous communication diminish rapidly with distance. It appears that after about 20 metres, communication has to be arranged in some manner for it to actually happen.

Spontaneous communication is however, an important part of innovation. Chance meetings, the communication of seemingly unimportant information, and the exchange of feelings via gestures and body language, are an important source of creative energy. If I encounter a sympathetic soul and present him or her with a current problem, I am likely to receive help, and may even receive hints and suggestions without having to ask for them.

☐ *Does the Innovation Cell have to be made up of volunteers?*

• Yes it does. Here is the simple reason. If you yourself are voluntarily making the leap to innovate, then your team members should have the same chance.

☐ *What is the complicated reason?!*

• The other reason has to do with remaining in the uncomfortable zone, and is as follows:

5.3 Remaining uncomfortable and agile

• When a team sets out on a journey of innovation they will come across many obstacles.

☐ *You mean they will have technical problems, go down dead-ends, and find themselves short of resources?*

- Yes, but in some sense there is a solution to all of those problems.

□ *But what if they get stuck completely!*

- Well even that can be resolved, by returning to the reasons for inviting them into the team in the first place.

□ *You mean the goal might need to be changed?*

- Exactly. So all these problems can be solved - except for one obstacle - disillusionment. If you, or a member of the team wants to give up, the innovation journey is in mortal danger.

□ *If someone doesn't want to continue, couldn't we simply raise the incentive?*

- Piling on the pressure or putting more jelly babies on the plate are methods that work well in the paradigm of "efficiency". But they tend to be counterproductive when it comes to "effectiveness".

□ *Do I have to learn new methods for motivation if I want to become an effective leader?*

- First you have to accept that you cannot motivate another person. You can only motivate yourself. Of course, you can entice, coerce and threaten- but motivate? No. Each individual has to motivate himself. What you can do is create an environment that enables an individual to be motivated. For this you will have to learn the art of war. You must anticipate the movements of opponents, ensure the morale of your troops, and make sure they are ready to fight. But more than any of these, you must help them have the courage to deal with failure.

Essay

Why autonomy?

You may ask yourself the question why should an innovation cell be an autonomous unit in the company? The answer is simple. The innovation cell is there to create potentially disruptive innovation. As you don't want any disruption to hold up the rest of the organisation, it is wise to keep the innovation cell separate. Let's look at the answer in more detail. When I give autonomy to a group I give it access to power and I also expect responsibility. The group will have to accept responsibility in order to honour the authority granted.

Simplicity of organisation

An autonomous unit is the simplest unit possible. With an autonomous unit the organisation goes "back to basics" and imposes less structure. The content is therefore able to develop more freely. This is important for innovation. Innovation can be stifled if it is overwhelmed by too many time constraints, too many rules or management interference. In some cases the answer to the innovator's problem is already present, but needs dedication to be recognised and freedom to be explored, before it can be shared with outsiders.

The innovation cell is a small and independent unit not unlike an embryonic company. It is free to find its own path and when successful, may inspire the rest of the organisation to follow course.

Trust in accountability

The innovation cell is autonomous when it has control over the significant resources needed for the innovation project. Typically, these will include the required experts, the project budget and the communication link to key suppliers and customers. The command of these resources enables the innovation cell to run the project without major interference or delay. By owning the resources, the innovation cell becomes accountable. The Controlling department will find it easier to check the use of resources.

To give an innovation cell autonomy is a sign of trust from management. By trusting the innovation cell, management gives the team an incentive to honour this trust. It is unlikely that the team or its members will waste the chance they are given. There will be group pressure on the team members to commit to the task. Like the sailors on the "Golden Hind", it will be a matter of honour for everyone on the team to keep going through thick and thin in order to reach the goal.

A fresh start through independence

It is in the nature of innovation that the successes of tomorrow can only be perceived in small and seemingly insignificant signals today. The trick therefore, lies in identifying those signs, preferably before they become common knowledge. The innovation cell has a twofold advantage: it is not plagued by the successes of the past and it is sufficiently independent from the rest of the organisation to try yet unconventional ideas.

A successful company is plagued by its past when it comes to predicting the future. The living memory of how successes were created inadvertently raises the expectation that the once so successful methods will work in the future. This makes the company blind to changes around it. It is also difficult to distinguish between the once successful methods and the many habits developed along the way. These habits will slow progress down and make the company inflexible.

The autonomous innovation cell is not burdened with the past. Its isolation from the organisation allows it to experiment without attracting the attention of the casual onlooker. It can therefore escape early judgement and any preconceived idea of what is right and wrong.

The autonomous innovation cell is a like a greenhouse for new ideas. It enjoys the protection of the cell. But just like any greenhouse, it will require weeding.

A 'tour de force' with speed of decisions

Long gone are the days when functional management dominated industries such as the automotive industry. Functional management proved itself to be too slow for the pace of industry. In particular, the arduous decision processes delayed development. Too many heads of departments were involved in the day-to-day decisions of development projects. As the development process became more and more standardised, the advantages of a functional organisation receded. The decision process was handed over to the project manager, a leader designated for a particular development project. Decisions could now be carried out more quickly, as fewer people needed to be consulted.

In innovation projects, the designated leader himself can become the next "bottleneck". An innovation project is never a standard process. The innovation project leaves the beaten track and enters uncharted territory. The project team needs all its wits and all its skills in order to lead the project to success. A designated project leader is likely to be over-challenged. The authorised project team, however, is not. It can decide to follow two paths for a while, backtrack when coming to a dead-end, or even decide to abandon the project and reconvene, should such drastic action be required, without ever losing face.

Aiming at success with dynamic failure tolerance

With autonomy the innovation cell becomes more tolerant to faults and possible failures. In particular, it can determine its own path and even re-define its goal. What would you say as a manager, if you asked for the product A to be developed and your engineer comes back with product B? You might not be pleased. If however, you asked your engineer to invent a new product A and he came back telling you that product A is a very difficult project, but product B looks promising, you might ask him to proceed.

Experience shows that innovation cells are particularly good at finding new paths when others have got stuck. The team creates new options and chooses which one to follow. When a given goal seems no longer attainable, the innovation cell will reassess and find a new goal worth pursuing. Sometimes the spin-offs thus created are more valuable than the initial aim.

☐ *That sounds complicated!*

• It's quite simple really. Just start with yourself. If you can deal with failure, your team will be able to as well.

Essay

What do I gain? – Path of Innovation leadership

"How is it done successfully?" has become the dominant question in our day. Conferences and presentations address the subject in "Best Practice" sessions; book shops and libraries cover the subject in their self-help, management advice and biography sections. The "How" is important, but we focus so much on it that we increasingly neglect its counterpart, the "What"! In fact "How" and "What" are like two sides of the same coin. In their interplay they reveal the profound impact one has on the other. Regarding innovation there are some important "What"-questions, which lead to new ways of "How". Among these are:

- The question "What opportunities do we neglect by focusing exclusively on efficiency?" results in the ability to see with new eyes

- Asking "What can we create together if we start with an open mind?" opens the possibility for new types of leadership and professionalism

- When we ask "What risks am I / are we willing to take?" we move towards the entrepreneurial spirit of innovation

- The question "What contribution may I be able to make?" enables us to rediscover our creative forces

These questions act like signposts on the path towards self-organisation and an entrepreneurial spirit in innovation. But it is not an easy journey. You have accepted responsibility for innovation, for a team, a product, technology or market. You are risking a lot by leaving the mainstream, sticking your head out into the unknown. You have bet your reputation on the success or failure of the team. But do not despair: if you genuinely strive to answer these questions, individual and collective rewards will be the natural outcome. Learning what it takes to become a true entrepreneur and leader will be your biggest reward.

Seeing with new eyes

What opportunities do we neglect by focusing on efficiency as the only yardstick for work and human life? The answers may be painful. As much as the efficiency paradigm makes many tasks of daily life easier, at the same time it diminishes the richness of being fully alive. Consider for example, that under this paradigm the most efficient performance of Beethoven's 9th symphony may be the one, which lasts only seconds. Absurd, you will say. But remember your last trip to a supermarket in the USA, buying for example strawberries. They provide a perfect example of just how far we have already entered the "efficiency" cage. Californian strawberries have been constantly selected for efficient shipping, efficient shelf life, efficient appearance and price. What they have lost on the way is their "strawberry-ness". It is easy to demonstrate - just taste them!

But there is hope. We can leave this ever more rapidly rotating efficiency wheel behind us if the results no longer serve our human needs. It is tough, but there are successful examples. One originates in Italy and has made its home worldwide on the Internet. It is called "Slow Food". The following is an excerpt from its web page: "The international movement was founded in Paris in 1989. The definition contained in its Manifesto conveys a very clear message: a movement for the protection of the right to taste. If Slow Food has grown into a large-scale international movement, with over 60,000 members in all five continents, it is precisely because the concept of "pleasure" is a complex one, encompassing many meanings and involving many aspects of our existence."

It can be useful to ask what do we lose out on in products such as automobiles, textiles, furniture and other material goods, because of the singular focus on efficient production? In our view, we increasingly lose genuine novelty in and attraction to these products. This is because they become increasingly look-alike, available for a similar price, and produced with the same quality (or lack of it). This satisfies the basic needs of many, but fails to satisfy the human desire for differentiation. There are approaches, which can turn the trend around, for example concepts such as flexible production, which use customisation of mass-produced products as the way out. Innovation is another very different approach, in particular when the innovation attempts to be disruptive.

The efficiency paradigm may diminish our capabilities for variation, creativity and liveliness. This is the natural consequence of a thinking pattern in which everything that impedes sequential flow, needs to be eliminated. It is effectively the old assembly-line thinking pattern that has been perfected to an extreme.

The world is changing

But nothing stays the same forever. Everything has a lifecycle, which includes its own "death". To create something new means we must notice when something else is at the end of its lifecycle. Working in an innovation cell trains you to spot these subtle changes in their early stages. The first result is that you notice more. But seeing without action is not enough. You too, need to change.

You need to change

When you notice more and become aware of the "efficiency" filter in front of your eyes, your horizon can broaden. Because you have already learned to act bravely despite your fears, fear will not blind or deafen you. Now that you see clearly, you can act accordingly. The necessary change required to do this successfully, includes letting go of the efficiency paradigm as the sole approach to all opportunities in life. You will learn to wait when waiting is appropriate, and to move ahead when that is needed. But most of all, you will develop a portfolio of skills far beyond those required by efficiency, including a sense for the appropriate application of these skills. The essence of the changes you must undergo was summarised more than 2000 years ago, in the writings of Tao Te Cheng: "Colours blind the eye; noises deafen the ear; flavours numb the sense of taste; thoughts weaken the mind; desire withers the heart - the Master observes the world, but trusts his inner vision. He allows things to come and go. His heart is as open as the sky."

Translated into our times it may read: efficiency lets you see, smell, hear and taste only a certain segment of the world around you. Awareness of this allows you to remove these blinkers, should they prove inappropriate or of no use. Having reached this state of mind you are now ready to become a true leader.

The possibility of new types of leadership and professionalism

What can we create together if we start with an open mind? An open mind and heart are the prerequisites for creative collaboration in the innovation adventure. An open mind lets you see the opportunities and obstacles that lie ahead of you; an open heart allows you to share yourself with others, so that they can accompany you on the journey. The biggest reward may turn out to be that you are no longer alone when it comes to making decisions. This does not mean that under certain circumstances you will be relieved of making difficult decisions for the group. But in situations where you are the one who must make the next move, the group will stand behind you every step of the way. Then having moved on, you will find that the whole group is following you.

What you gain

None of this comes for free. You have learned to see with new eyes; you are opening up to the power of collective creation; you see new and exciting opportunities before you. But you also become aware that the transformation of these opportunities will transform you as well, whether you are the formal leader of an innovation cell or one of its professional members. These transformations are very attractive. As a leader you will gain new skills for your leadership portfolio. As well as the classical leadership approaches of telling, selling, testing and consulting, you will enter the realm of true co-creation, gaining first hand experience of the freedom of creating together the direction you want to take. This includes letting everyone speak their mind, having the conversations that matter, and being able to count on each and every member to be responsible for him- or herself. As a leader you will learn and become comfortable with the fact that leading includes the following:

As a professional, you will expand your technical skills into the field of leadership and collaboration. The biggest gain will probably be the opportunity to leave the cage of specialisation, and rejoin the more natural interdisciplinary community. Whereas in the past you functioned in a somewhat fragmented way, you will now be able to function holistically. One immediate effect will be the realisation that one plus one really does make more than two. The price paid is your willingness to take on full responsibility for all aspects of your work, not just the technological part. The non-technological aspects were really already a part of your work, but had been lost because of the hierarchical nature of the organisation, and the separation of work that goes hand in hand with specialisation. It is a rewarding price to pay, because it will return to you a significant part of your humanity and dignity.

At this point it becomes clear that everyone involved in an innovation cell will find himself or herself standing at a significant crossroads early on. You need to make a wholehearted decision - are you willing to shoulder these additional skills and responsibilities?

Being different does not mean being separate

When you work as part of an innovation cell you start to realise what different skills the people around you have. Some of these may overlap with your own; others could be complementary. You will also experience which values we automatically associate with different skills. In our society certain values are more appreciated than others, so that some skills are seen as less valuable, despite their objective importance for the success of a project. The over-valuing of hard skills, such as technology, finance or marketing, and the under-valuing of soft skills, such as openness, clarity, approachability or empathy, provide ample examples. But different skills seen from an integrated perspective are just that – different skills. For successful innovation, the presence and appropriate application of many different skills is required. Attentiveness, for example, is as important as technical competence with computers.

This clearly poses a challenge for everyone in today's work environments, because status, income and power are assigned to certain skills. We are automatically geared towards judging and separation, whereas the diverse skills needed for innovation come together in the innovation cell. Here again, you will need to be open to new ways of thinking and acting. The people in an innovation cell are not equal, because each has different skills. But at the same time no one is superior because of his or her particular skills. Being different does not lead to separation in innovation cells. Instead, an old value is rediscovered: genuine acceptance of team members based on mutual respect for each other. To gain this, everyone must work hard on him- or herself.

Soaring productivity

Mutual respect provides benefits not only for individual members of the team, but also in terms of economic results. It is a well-known fact that people who feel appreciated and respected for their contributions, tend to create better results. It seems obvious: self-realisation through innovative work, results in more innovative products.

At this point you will have achieved true alignment and commitment, something business leaders around the world desperately seek, yet often never find. Alignment and commitment depend on the maximum involvement of everyone in the group. The leader's open heart is a prerequisite for creating communities where everyone feels welcome and appreciated. This too is true for innovation cells.

Being an entrepreneur of Innovation

What risks am I - or are we - willing to take? When King Arthur and his knights set out on the quest for the Holy Grail, each one chose to enter the forest ahead of him alone and set off on his own unique path through the forest. Entrepreneurial spirit is very much like this. The entrepreneur of innovation is willing to find his or her unique path to "invention". It is a path no one else has taken before. As in the case of King Arthur's knights, this is a dangerous undertaking, because one could lose one's "life". The risks today differ from those encountered in the quest for the Holy Grail, but their challenging nature is the same. Instead of mighty dragons, we may face an investment banker on the brink of withdrawing necessary funds. We may not get lost in the dark forest, roaming around forever in circles, but we might enter into research activities that after many years of hard work prove fruitless. The path we take is dangerous. The successful innovation entrepreneur needs closely to examine his or her tolerance for risk and failure, because these will be his constant companions.

When this examination is over, and despite it you want to move forward, there will be much to gain. One of the rewards for the leader is an increasing confidence in the natural flow of things. Here is another part of the Tao Te Cheng which expresses this idea very well: "...The journey of a thousand miles starts right under your feet. Rushing into action, you fail. In the attempt to grasp things, you lose them. Forcing a project to completion, you ruin what was almost mature. Therefore the Master takes action by letting things run their course..."

When this sense for the "natural flow of things" is matched with clarity of mind, openness of heart and a strong will, the entrepreneurial journey will not only be successful, but also enjoyable. And enjoying the journey itself might be worthwhile in itself.

Example

The power of the right idea

Imagine the following: you are Joe, the CEO of a small but respected engineering software firm. A year ago you won the largest contract ever struck in the history of the company. The contract was your entry to the premier segment of the market. You made considerable promises to your client, CarCompany, to win it, among them the development of a key interface, enabling the two product lines purchased by your client, to work seamlessly together. The interface was at the core of the negotiations, and its availability eighteen months after signing the contract was a critical item in the contract.

Immediately after the contract was signed, a development team was implemented to create the promised interface. This team consisted of 60 developers and their managers, and has been in place for exactly 12 months. The delivery of the interface is just 6 months away.

Ed, Vice-President of Product Development has just left your office. He told you that the delivery of the interface is in jeopardy, and that the team has not yet even started developing code. All your dreams for your own and the company's future are collapsing around you. The renewal of your CEO contract depends on a successful implementation at CarCompany. You lose all trust in the development department - this is the second time they have let you down. You do not know what to do - and for a moment a feeling of panic races round your body. You sweat and your heart rate rises to levels way above acceptable limits. You make an effort to breathe slowly, and then remember that there is a person in your organisation who had successfully turned round a similar situation. You pick up the phone and ask him to come to your office right away:

Joe: I have just been told that the interface product for CarCompany can't be delivered as promised. If we cannot correct this situation, they will terminate the contract. If this happens it will harm the company greatly, and the company might not survive it financially. In your previous assignment you demonstrated quite some talent turning round a seemingly hopeless situation. Do you think you can help again?

Mark: Yes, I think I can. Rumours about this are already afoot. I know just how bad the situation is from the project manager. But there might be a chance of turning it round. It will require a very un-usual approach; one we are not really used to!

Joe: OK, tell me what is needed, and how I can help to get it done.

Mark: In a nutshell, I need your approval for the following changes. The project team will be physically relocated on one floor of our de-velopment building. The team can structure the workspace in any way necessary for productive work. We require CarCompany em-ployees to participate in the team, especially during the next 2 months. For the next six months, until the project is finished, the team will operate outside our standard project management proc-esses. We will provide you with a weekly update on progress, but whatever happens, you must promise not to interfere. We will use self-organisation and a strict weekly feedback method to keep us on track. We estimate a 50/50 chance of pulling this off. If we fail, you can fire us. Supposing we do succeed, you will reward the whole team publicly, according to each team member's contribu-tion.

Joe: This is most unusual, but I suppose we might succeed this way. I trust the team will do everything necessary to turn the situation round. Go ahead, you have my approval to work as you propose.

The product was delivered to CarCompany one week before the dead-line. It had a quality rating double that of the normal BSWC products' rat-ing at this quality gate. The team achieved this remarkable result primarily as follows: project management was split into two roles, that of technical project manager and a "political" project manager. The latter was Mark's responsibility. His role was primarily that of a coach encouraging the team to do what it already excelled at, as well as defending it from the many panicky interferences of the executive team. These visited the team control and collaboration centre on a weekly basis for reassurance, but were often confronted with unpredictability and creative forms of chaos.

The technical project manager's role was to focus the team on understanding CarCompany's requirements, and on developing the software solution. Despite being under pressure, the team took the time during joint meetings with CarCompany, to pinpoint what it was the customer really wanted; team members applied a radical feedback process which allowed them to know exactly how their individual contributions were helping in the achievement of project goals; a manual, simple but very clear project status method was applied, which everybody could interpret directly; and the team organised itself dynamically according to its and the customer's actual needs.

Rediscovering our creative forces

What contribution can I make? We often limit the richness of our contributions by the way we think about ourselves. Our education, our daily routines and many other influences contribute to this situation. Perhaps we think we cannot sing, because when we were 8 the music teacher told us so, and consequently we never tried again. The same may be true of any of our creative capabilities. After a while we see ourselves only in ways others have directed us to see. Combining this with the tragic illogical ethos of the work place, where conformity buys us safety, we cripple our souls to wastelands, where creativity, enthusiasm and curiosity linger on like parched crops. But when we engage in innovation cells it is like watering our soul. What was buried and seemed to have died has a chance to flourish again. We become more alive in many areas of our life; areas we may have forgotten existed. According to Jill Badonsky the nourishment for our creativity that we need to rediscover includes:

- Bringing our emotions back into awareness
- Thinking outside the box, acting differently and seeing the world around us with new eyes
- Bringing play, fun and laughter back to the innovation work-place
- Replacing discouragement with support and engagement
- Losing the fear of making mistakes, giving up addiction to perfection, enjoying both process and results equally

- Trusting one's intuition and being true to oneself, being liberated from the opinion of others

- Opening up space for celebration and letting go

- Releasing frustration and anger as a force to deepen our creativity

- Becoming aware that "continuing despite set-backs" is part of being creative

When we work consciously with these items, we discover our true potential for creativity. In innovation projects we need an abundance of these capabilities because they are actually the only real resources we ourselves completely control. Embracing these principles brings rewards not only in our work, but in all the other parts of our life as well.

Example

Transcending the fear of failure

This is a very personal example. It is about a moment in George's life, when he thought his dream of building his own, independent consulting practice would come to an abrupt halt. George was in his third year as an independent consultant, after having spent thirty years in corporate life. He had committed all his personal resources to building-up the business. Since starting as an independent consultant, he had experienced successes as well as some setbacks, but overall the business had grown steadily for over two years. The third year looked promising too. Things had started well, but after three months of business, began to slow down somewhat. Although he had many interesting engagements, few of them generated revenue. After four months of little income, his short-term financial resources became stretched. The situation was worsened by the general slowdown in many of the markets he served. His clients suffered in the same way, which resulted in ever-increasing delays in the payment of invoices.

At the beginning of the autumn the situation had developed to a point where the ebb of cash flow became threatening to the well being of his family. These were some of the choices available to him:

- Give up the dream of independence and join a larger consulting company
- Give up consulting altogether and go back into the corporate world
- Continue the chosen path and liquidise all long-term financial assets, thus providing another 12 months of financial survival

He was 55 years old now, and finding further employment in his previous environment or with a consulting firm seemed unlikely. Selling all long-term assets would provide relief for a year, but if business had not improved by then, the situation would be even worse. Even if business did improve, selling everything would mean significant financial penalties, because of the structure of his investments.

During one particular night, after waking up in a sweat from a horrible nightmare, the following internal dialogue took place in his mind:

George's shadow: You fool. This is it. You will lose everything, just because you wanted to follow an impossible dream. Nobody pays real money for your idea that work can be effective and enjoyable at the same time.

George: But it *has* worked for a quite a while now. Remember the great research workshops with GlobalCorp and the joy people expressed during them. What a delight it was to be a part of that. I will never forget the energy and enthusiasm that was present in those events.

George's shadow: But that was kid's stuff and people only spend money on that sort of thing when times are good. Just look around you - unemployment is on the up and your age is a showstopper for employment anyway.

George: Yes, I know. I feel empty, and it is difficult to see a solution. It seems as if everything is falling apart.

George's shadow: Yeah! Like I said. You are a fool. You will end up on the streets with no money, doomed to live off social security. It might be OK if it was only *your* life we are talking about here, but you are going to ruin your wife's life too, damn you.

George: Wait a minute. I didn't start this for the money in the first place. During our 30 years together, Marie has always stood behind me. She has given me so much encouragement. When I decided to become a consultant, she supported me whole-heartedly. How can I let her down now? What is the worst-case scenario?

George's shadow: You and she will end up on the streets and then...

George: Shut-up. Nobody will end up on the streets. There are only a few more years until my retirement payments start. I will find a way to bridge the gap. Anyway, more importantly, I love what I do. I'm finally following my most glorious dream. How could I give it up just because I'm afraid of losing my financial security? There is so much, which can be done. If the situation does not improve within a week, I will sell some of my assets. There is always a chance that things will get better. I just have to believe this.

After this internal dialogue, George slept well until the next morning. He woke up full of energy. During breakfast he talked about his ideas with Marie, and she encouraged him to stick to his decision. That very afternoon he received the following email: "Dear George, we have a client who needs help urgently. We can't do the job ourselves, but you would fit the bill perfectly. Are you available? It will be a very well paid job. Please call us today."

George is now in his fourth year as a successful consultant. He is booked out for several months in advance. He loves his work more than ever, and it seems that this aspect particularly, is what attracts clients. But he has not forgotten the night when all seemed bleak and hopeless. Facing his biggest fears seemed to unleash an energy source he was not aware even existed.

6 Selling innovations

- Let's take a moment to look at our achievements so far. We have come a long way. Like Columbus, we have left our safe harbour, made our way across the big water, and find ourselves on a new shore!

□ *Hooray!*

- Don't rejoice too quickly! There is one niggling little question: "Are we in India?"

□ *Who cares?*

- Well, certainly the Duke who paid for Columbus' journey does. Having found America, Columbus faced his biggest problem. He had been supposed to find the route to India. Instead, he came back having found a large and seemingly useless piece of land, which was sitting there blocking the spice route!

□ *But America is a great asset!*

- Yes - it is now. But in those days it wasn't. It had to be turned into an asset.

☐ *How do you turn an innovation into an asset? Can't you just sell it?*

Essay

What creates sustainability?

Innovation is an important part of companies today for the following two reasons: firstly innovation generates new products and services; secondly innovation renews a company continuously. We have looked at the first reason in detail. Let's look now at the second. We will use the innovation cell as an example of the continual renewal. We know that innovation cells can create new skills, technologies and competencies. In the following paragraph we will investigate how innovation cells create learning organisations and redirect the culture of companies towards innovation.

Interface to organisation

The innovation cell is created within the company, and its members have to return to the company. Immediately before starting an innovation cell the following ingredients should be available:

Innovation cell starter kit:
1. a rough idea of the innovation to be achieved
2. people who are prepared to venture forth and explore
3. resources to fund the exploration
4. the will to focus effort

The innovation cell can then be formed and work can commence. Like any other project team, it draws its resources from the company and its various departments. Unlike the usual project team however, the innovation cell demands full authority over use of resources. This means that for the duration of the innovation cell, the members are fully dedicated to the innovation cell, and don't serve any other taskmaster. This is a difficult lesson for many companies to learn, where the best people traditionally carry out a multitude of tasks, and where managers still act as if they believe they own the people who are working for them.

Of course the innovation cell members return to their places in the organisation. This is what the innovation cell then returns:

Innovation Cell heritage

1. a product and technology developed to the point where the standard development process can take over

2. people with the know-how and skill for the new product and technology

3. the will to bring product and technology onto the market.

When the innovation cell is disbanded and the project comes to fruition, their former colleagues can rejoice. The best people have returned and once again they are in full charge of their resources. They may however, still feel a little hurt about the priority treatment the innovation cell received. It is your job as a manager to help them feel secure in the knowledge of their own importance.

Example

Personal courage leads to business success

In the early days of the Computer-Aided Design and Manufacturing business (CAD and CAM), the market was swamped by a myriad of products. Each niche had its own CAD system. MyCorp was one of the first companies to question whether this collection of individual solutions was really helpful for a company's overall product development process. Therefore, as early as 1980, MyCorp started to create a portfolio of selected CAD solutions which was intended to cover all technological aspects of computer aided design and engineering. The portfolio consisted primarily of three cornerstones: two CAD systems, one providing two- and the other three-dimensional capabilities, and a computer aided engineering system with capabilities for finite element analysis and stress calculations. MyCorp's idea was to connect these systems in such a way, that whole product development solutions could be proffered instead of just isolated design and engineering tools.

But after a few years it became clear that the partners, whose products MyCorp wanted to connect through interfaces, extended their products into each other's technology domains themselves. Soon MyCorp's portfolio started to look like a collection of competing products rather than a well-integrated suite of interrelated products. The situation caused MyCorp some internal pain. For some time two basic business ideas competed with each other internally: to stick to the idea of complementary partner products or to select one product as the lead solution and to disband the others.

At a large computer and technology fair in 1984, a small group of people who were convinced of the long-term market potential of one of My-

Corp's partner's three-dimensional solution, decided to take charge. In preparation for the fair, marketing material had to be prepared, and the group decided to put all its eggs into the three-dimensional basket. A poster was designed with the help of an external marketing agency entitled "Three-Dee, the computer aided design Bonbon from MyCorp". Except for this group and the external marketing agency, nobody knew about it. Four thousand posters were printed and shipped secretly to the fair. There they were displayed at strategic places around the MyCorp booth. The poster created an éclat the moment the show opened. Representatives from My-Corp's partner, which provided the other design system, together with their internal allies, complained instantly. They saw the poster as a sign that MyCorp was diverging from its business idea of a balanced computer aided design product portfolio. MyCorp senior management could prove that it had no involvement in this coup. The culprits were easily identified. The following represents the dialogue between the manager who was responsible for the poster and the national manager responsible for MyCorp's engineering product portfolio:

Frank: You must be crazy to create such an éclat. The national manager of
          the design product is fuming because you openly dismissed his
          product with your poster. You didn't even bother to inform him -
          or me. We should have talked this over together.

George: I doubt it. You have probably forgotten how many times my group
          has tried to make you aware of the increasing similarities among
          our computer aided design products, and how this has confused
          our customers. We needed to take a stand.

Frank (shouting): Do not lecture me; I'm still your boss. You know that
          your action could lead to your immediate dismissal from MyCorp.

George (a bit smaller and paler): I didn't mean to lecture you. But you have
          told me often enough that you yourself don't understand the corpo-
          rate strategy regarding the engineering market. In the end we will
          be measured by what we sell, and how satisfied our customers are.
          I firmly believe that my group's action will help MyCorp to be
          more successful in this market.

Frank (less angry): You are right about my personal view, but this does not
          justify your group's action, which makes senior management look
          foolish because they were ignorant of it. You will have to take the
          consequences.

Similar conversations were held on several more occasions, but the tone of them slowly changed. During the computer fair, several significant My-Corp customers had come to Frank and congratulated him on his clear decision to select the three-dimensional design system as the lead MyCorp computer aided design product. One of these visitors had been the head of product development at a large automotive company. Higher MyCorp management could not ignore these customer statements of support.

After the fair, the team that had created the poster experienced both support for its courageous action and blame for the way they had carried it out. But after some months it became clear that it had been the right decision. Within months the leading companies of a significant European product industry declared the design system, which had been featured in the poster, as its product design standard. Letters were sent to suppliers and partners, encouraging them to follow these companies' lead, and select this system as the main design system. Over a period of twenty years it became the leading global standard in product design.

Some of this success can probably be traced back to the decision of a small group of people who had the courage to follow their convictions.

Co-dependence Innovation Cell and production

Innovation and production can be likened to oil and vinegar. You have to stir them occasionally. Production without innovation will lead to old and outdated products and technologies; innovation without production will not make any money. In other words, production and innovation are co-dependent.

Innovation is an undertaking that needs resources, in particular financial resources. An innovation cell is an investment in the future of production. If possible it is wise to use some of the earnings from production to finance the innovation cell. The financial link will create a bond between the innovation cell and production, and generate an incentive to create new values for production.

Production on the other hand, has to rely on technologies to run its processes. Production will become outdated if it neglects to update its technologies. In many cases the new technology sought after in production will also create new opportunities with products. Therefore the most natural task for an innovation cell is to develop a product on the basis of a new technology that production is seeking.

Your job as a manager is to ensure that the production department's desire for new technologies and your customer's expectation of new products reach culmination in an innovation cell.

Pipeline

You may ask yourself the question "How many innovation cells does my company need?" The answer is "probably more than one". Let's work it out together.

Example

How many Innovation Cells are needed?

Innovation projects may fail. Despite this risk, your job as a manager is to ensure that your people don't get branded as failures, and that enough innovation reaches the customer. Let's listen in on how Bill, responsible for advanced engineering, and Bernie, in charge of R&R at MyCorp, arrange their 'innovation machine':

Bill: With the experience of our first three innovation cells we should now be able to arrange a stable state process for the delivery of innovation to our customer.

Bernie: OK, let 's look at the figures. We had 3 innovation cells with a total budget of 1 million, accumulated project time of 25 months and a success rate of 63%.

Bill: A sample of 3 innovation cells is probably not statistically significant.

Bernie: It doesn't have to be. We can always adjust the figures when we know better. Now my question- if we want to deliver two big innovations every year, how many innovation cells do we need.

Bill: By doing the maths, I get the following figures: We need 2.4 innovation cells in parallel, each occupying 6.6 staff. That is an annual budget of 800 thousand and 8 people.

Bernie: In other words we can offer top management two decent innovations every year with a budget of 800,000 and using 8 people.

Bill: I am afraid that is not all. The innovation pipeline has to be filled as well.

Bernie: How could I forget? Let's do our calculation for budget and resources for the whole innovation pipeline.

Innovation can be made into a fairly stable delivery process. With enough innovations in the pipeline, the success or failure of each individual innovation is no longer critical. If one innovation proves to be impossible to achieve for the time being, another one is just round the corner. No one needs to get hurt for accepting a risky assignment; no customer needs to be left disappointed.

The innovation cell is a way to work through a complex project for the renewal of product, technology and in the end of the company itself. So first ask yourself "How many new products can I introduce into the market?" or "How many surprises do I want to create for my customers?" Let's say I want to launch 2 new products a year.

Now I have to ask myself: "what is the risk associated with a typical innovation project?" or " how likely is it for an innovation cell to fail?" Let's say the innovation cell typically has a 50% chance of succeeding. In other words, I will need 4 innovation cells per year.

Now I check how many resources and how much time are needed by asking the question "How long does the typical innovation cell last?" or "How many people and how much money do I have to put aside for a typical innovation cell?" Let's say a typical innovation cell lasts six months and takes the work of three full-time members. Then I will have to plan for 2 innovation cells running all year round. As for the budget, I will have to plan for resources and the work of six people all year round. Please note that at this time one may wish to readjust the initial hypothesis, depending on how much innovation your company can really afford.

I may choose to dedicate 2 rooms for the innovation cells, and place them in a strategic position in the company for visibility and attention. Please notice that I have created a pipeline for innovation: a highly complex innovation project will now be dealt with in a manner that will no longer disrupt the day-to-day product development and project management processes. I have also created an incentive for the company to articulate innovation in a manner visible to my company's customers and suppliers. The visibility will initially earn you "Brownie points". As you persevere and continue to feed the innovation pipeline, you will notice your customer credit- and bank-rating going up.

Example

Top management attention nurtures the Innovation Cell

Just as a shop benefits from other good shops in the neighbourhood, so do innovation cells benefit from other good innovation cells in their neighbourhood. Imagine the following situation: Richard, the president of MyCorp, likes to know what is happening in the company. He keeps key resources such as finance, close to his office, so that he gets up to date information and early warning about when his attention is required. Bernie, in charge of R&D, likes to keep the innovation cells in a row in the research and test centre, and he moves his office right next to them.

Richard: What 's new in R&D, Bernie?

Bernie: I am happy to give you a quick briefing here. I would be happier still, if I could show you what's new. A picture says more than a thousand words and an innovation cell displays more than a thousand power-point presentations.

Richard: OK, I 'll come. I'm going out for lunch with Lee, the VP from Samiol. Do you think I could bring him along?

Bernie: One of the innovation cells is working on an innovation for Samiol. The second innovation cell is making great progress, but is still looking for a key customer.

Richard: I'll bring Lee along after lunch. You might like to warn the teams. Lee can't stand empty cardboard boxes. Make sure they are out of sight.

Life in an innovation cell is dynamic. In MyCorp the president knows that the act of creation is rarely attractive or tidy. He is happy to take a stranger into an innovation cell, and confident that the activity going on before their very eyes will inspire faith in the ability to innovate. Bernie knows that a visit from the company president will encourage the members of the innovation cell.

New Ventures

Now for the ultimate and strategic question: "Why *make* innovation if you can *buy* it?" After all, there are start-up companies, which - as luck may have it - just offer innovation to you without you having to do it for yourself. The answer is simple and two-fold. If you find a start-up company that fits your bill, then go for it. Otherwise use the innovation cell as a start-up company within your company.

The new venture as a start-up company setting up in the market has to make do without a lot of things the innovation cell has already built-in. The innovation cell has a ready-made environment for its product and technology. It consists of people proven and trusted in the company. If the innovation cell fails, its members are not doomed. In other words an innovation cell is quick to set-up. It can be treated like a venture company within your own company, without all the red tape, and with none of the downside risks attached to a venture in the open market.

A venture company and an innovation cell may have a number of things in common. Both owe their existence to an ambitious goal, such as the implementation of a new technology, or the launch of a new and revolutionary product or business. It takes determined pioneers both to start a venture company and to run an innovation cell. After the initial period, and as the venture matures, the initial pioneers may no longer be the right people to run it. This may well be the most trying time for a venture company. It is no problem for the innovation cell, where the company surrounding it can take over the initiative, and the innovation cell can "retire gracefully" taking with them all the pride belonging to a successful launch.

Unlike a venture company, you can use an innovation cell to launch a new business or technological platform. For example, you have used the first innovation cell to create a novel product with a new technology. Why not build on the initial success and launch a set of products with the same technology, thereby creating a new business unit for production and marketing?

Example

The first step starts the game – the second perfects it

When I bake waffles it is usually not the first, but the second one that comes out deliciously crisp and brown. The first one may not be a success. Innovation projects seem to have one thing in common with waffles: it is definitely worth continuing beyond the first attempt. The following example shows that the technical solution of the first generation product is surpassed by the second generation. Richard, the design engineer for the first generation products, delights in making the second generation even simpler:

Bernie: What's it like being in the LawnBot innovation cell the second time round? Wouldn't you rather be out designing something novel, rather than rehashing the last generation you finished only six months ago?

Richard: Funny you should ask. You know I like both. But it is a heaven sent opportunity to be allowed to correct all the mistakes we built into the first generation, and get paid for it.

Bernie: Surely there are no mistakes in the first generation?

Richard: Let me put it this way: LawnBot did not have a track record before we came out with it. Our customer may have had faith in us, but he certainly did not have faith in the technology. We had to implement extra safety measures to gain acceptance. You remember we were also under considerable time pressure?

Bernie: If you were our competitor, how would you design LawnBot?

Richard: It could be a lot simpler. I'll show you ...

Doing a good job twice need not be boring. Break-through products in particular, tend to benefit from an early revision. It is advisable to carry out the revision before the notorious 'fast follower' company does it for you!

It is also possible to use the innovation cell to try, test and anchor a new cooperation. The innovation cell could even belong to both partners. If successful, a joint venture company may result. The innovation cell lends itself to many jobs in and around your company where new paths or methods have to be found.

Example

Innovation Cells bridge the gap among organisations

In MyCorp it has been particularly helpful to involve students in innovation projects. A student in an innovation cell can, for example, ask questions which an employee would hesitate to put forward. But that of course, is not all. For the student, the involvement in industrial projects is a safe way to test his abilities on "real life" problems.

MyCorp used the experienced gained from innovation cells to start My-LAB, a public-private partnership with Myland University. It has similarities to an innovation cell. In particular, it is cooperation across organisations. It also contains a number of individual disciplines, is temporary, and will last only as long as it benefits both partners.

The following dialogue illustrates the perplexed reaction MyLAB receives from a more staid member of society:

Steven: Welcome to the MyLAB, Richard.

Richard: It took me ages to find you. I just couldn't believe that a university institute chose a location right in the middle of a production company.

Steven: Well I'm glad you found us. The MyLAB is what is called a 'Public-Private Partnership', and not an institute of the university.

Richard: And what is a 'Public-Private Partnership'?

Steven: It is a co-operation in the true sense of the word, between a public institution, in this case Myland University, and the private partner, in this case MyCorp.

Richard: Does the MyLAB exist as a legal entity?

Steven: No. Everyone who works here has either a contract with MyCorp or with Myland University.

Richard: Who then, pays for the MyLAB, and who gets the benefit?

Steven: The infrastructure of the MyLAB is paid for jointly. The project finance is arranged according to the expected benefits.

Richard: Won't one partner try to outdo the other?

Steven: If that happens, it might be the end of the co-operation. MyLAB is built on the trust existing between the partners.

The MyLAB, like the innovation cell is a co-operation based on trust. It is the temporary union of trusting partners with joint goals, to create new products, new technologies and new knowledge for each other.

☐ *How do you turn an innovation into an asset? Can't you just sell it?*

● Selling the innovation is exactly what you will have to do, and it might be the hardest part of the job so far...

Example

As a manager you cannot delegate responsibility

The reasons for the failure of an innovation can be manifold. In the following example, a structured market approach was missing. The customer is particularly disappointed with the sudden demise of a promising and innovative product, as Michael in charge of advanced design at EuroCom freely announces to his supplier:

Michael: What happened to the AirWire? I spent so much of my time and effort persuading the platform manager to open the designs to accommodate the AirWire, and now this!

Leonard: I know how much effort we both put in. I thought we were still on track. What has happened?

Michael: You know EuroCom is very happy with the technical side of the project. Ever since you have been involved and giving us support through the innovation cell, we know that we can rely on MyCorp. What completely threw us, however, was the reply our purchasing people received from your sales organisation.

Leonard: We try to keep sales involved with the development in the innovation cell as best we can.

Michael: Maybe your sales people were too afraid to show you the quote they were about to send us. What they sent us is outrageous. They expect EuroCom to bear the full cost of the technology and product development. You can imagine what our purchasing did with the offer! All our programmes with AirWire have been cancelled.

Leonard: Is there any chance of revoking the decision?

Michael: It sounded absolutely final to me. It's a shame. I was beginning to enjoy my work with MyCorp.

The innovation cell in itself is no guarantee for success. The management responsible for setting up the innovation cell also remains responsible for the project. In the above case, the management failed to include the sales task sufficiently in the project. As a result, the technical solution was a masterpiece, while the commercial offer to the customer was a disaster.

6.1 Allowing things to happen

- First you have to allow things to happen. You may not be able to predict the outcome of innovation, and sometimes the outcome will be very unfamiliar, and very unexpected. In some cases you will simply have to admit defeat.

- □ *But I can't do that! Everyone will laugh at me. All the "lemmings" I was so eager to leave, will turn round and point at me saying, "Bernie always has to find his own cliff". In addition my boss and my clients may tell me it is of no use - how do I convince them that my idea is a good one?*

- Well - we have dealt with that particular fear already. Your innovation team has found something that neither you nor your team expected to find. You as the leader, see its potential. Now it is time to think about the positive side of innovation and how to "spread the word".

- □ *When I remind myself to "think positive", then allowing things to happen means good things can emerge.*

- Right. So its just like letting things grow. There will be weeds as well as corn. Pull up the weeds and leave the corn to grow – it's simple!

☐ *Oh - and don't forget to sell the corn, when it is ripe!*

6.2 Continue with efficiency

- After first allowing things to happen, the second task is to reintegrate the method of innovation into the existing organisation.

☐ *What a pity- I was just beginning to enjoy myself!*

- And so you should! You deserve it.

☐ *Thanks!*

- Having taken the risk of leaving the beaten track, you might consider making a track for others to follow. You and the team are setting an example for success achieved by unusual means. Some people may follow your example independently, others may hesitate.

☐ *I don't fancy this selfless task. Don't you remember the resistance the others put up? They don't deserve my help. Why should I help them?*

- I was thinking more about you and the help you will need. Your life will be easier if you let others share in your success.

□ *OK, I grudgingly concede this argument, but I am still not completely convinced. What can the existing organisation gain from my work?*

● There is a stigma attached to innovation, which goes by the name of "fear of failure".

□ *I know it well.*

● We both know that in order to innovate, it is essential to face this fear. If we could somehow make failure an accepted step on the way to innovation, we would have effectively removed the fear.

□ *How do we do that?*

● Don't you see that you just did it with the innovation cell? You created a new organisation dedicated to innovation, but allowing for failure! Because it was not within the old organisation, it didn't create a direct conflict there. We used the best elements of the old system. Do you remember where we started? The conventional system could not afford failure. By localising failure within the frame of the Innovation Cell, we can combine both systems safely.

□ *Hang on! We don't want to go back to the old way of organising!*

● No, we won't do that. But we will create a system where the old and the new can coexist.

☐ *I see! You want a system where both effectiveness and efficiency can exist! Great! I am with you.*

6.3 Remaining true to yourself

• Once you have allowed your perspective to change, innovation becomes simple. You can allow the creation of novel ideas, remove the fear of failure and be proud of your successes. By segregating the creative from the repetitive process, the conflict within the organisation is resolved.

☐ *But it isn't that easy!*

• No, you are right there. It may be simple, but it is not easy. For one thing, it requires a new form of leadership.

☐ *What happens to all the managers?*

• Some of them will want to be leaders; others will be more comfortable remaining managers. But then, as we discovered - management is still needed as much as it was before. It was the leadership for innovation that had gone missing along the way.

☐ *I wonder why...*

Essay

The end is the beginning

Supposing you have finally succeeded, and your innovation cell created the break-through you'd hoped for. Now you think the job is finished. You and your team received your well-deserved rewards, and celebrated. But you realise that something is missing. To start with it is just a faint feeling, but it is increasingly bothering you. Then it hits you like a rock: you have arrived *only to start all over again*. Your biggest realisation is that you like it; it seems to you a bit like surfing. But instead of just catching the next project-wave, you decide to think more deeply about what is happening. You realise that you are travelling along a path that looks like a band of S-shaped life-cycle curves. Whenever you reach the end of one curve you have to make the decision to jump to the beginning of the next. But even when you travel along one of the curves, there are several forks in the road that require your utmost attention. Let us pause for a moment and summarise the challenges you've faced:

- The start of your journey: you realised that the old ways of doing business were not good enough any more

- The big frustration: the efficiency paradigm does not work well for innovation

- A major chasm to cross: your fears, your courage and a big leap of faith in innovation

- Deciding on the path that seemed most dangerous: your first steps towards creating an innovation cell

- Personal growth: you befriended the unfamiliar and trusted your creative energy

- Enjoying the ride: the trust in your team and yourself is rewarded, and your first break-through innovation takes shape

What lies ahead of you requires yet another change of gear. It is time now to think about how to transform your breakthrough innovation into a major product in the market. It is necessary to lean over towards efficiency again. Now is the time to remember the successful rules and methods of efficiency- faster, better, and cheaper. You are at the beginning of one specific life cycle curve: the one related to the innovation the innovation cell just came up with. But this journey too, will not be easy. Along this path, until the curve comes to an end when your product has fulfilled its market purpose, you will encounter at least three more forks in the road:

- Crossing the divide from technological gimmick to marketable product
- Changing gear when your product has reached its market pinnacle
- Preparing yourself for your product's death in its markets

The further you go along this road, the more you need to be aware of the possibility that someone else will already start a development which will lead to the replacement of the product you are just promoting. Wouldn't it be better if the "someone else" were you? Thus the whole cycle starts again.

Innovation Cell is one answer to disruptive environments

Your option then is to use the concept of innovation cells as an on-going mechanism to challenge your own products. This might be a much better alternative to being challenged by somebody else, who might potentially, take your markets away from you. But you realise that the required continuous switching between effective innovation and efficient production asks too much of one person. It may not be possible, on your own, to disrupt what is known to work well, in order to create something novel and superior. You are going to need partners. Ideally the dichotomy of efficiency and effectiveness is already built into your organisational design. There are several ways of doing this. You can, for example, develop mechanisms for "spin-offs", provide seed money for Start-ups, or engage in Joint Ventures. The core idea is to separate the disruptive activities that are searching for something new, from those geared towards continuously improving the chances of something already on the market. If this separation is not made, a conflict of interests will result. It will express itself in endless fights for resources, a splintering of the organisational energy and confusion among your leaders and followers.

Innovation cells are one way of dealing with disruptive environments and innovation. Although the specifics of innovation cells are important, the generic concepts and methods behind them are even more essential:

- A clear understanding of the different, partly incompatible structures underlying creation and managing
- Understanding and determination on the part of leadership that innovation needs approaches beyond the "efficient assembly line" paradigm
- The trust that people will find the best solution for a compelling goal if left alone as much as possible

- Understanding of the interdependent and perpetual nature of innovation (creation) and production
- The realisation that the quest never ends.

Example

Success despite failure

Good strategies and efforts for change are sometimes jeopardised because of internal power struggles and rivalries among executives. Quite often this leads to failure in implementing what was agreed upon. But sometimes an idea is so strong and convincing that it cannot be ignored. Although it might not be acknowledged officially, it still becomes the guiding principle of an organisation's effort for change. Here's an example:

The strategic team had worked for a year to develop a new company product strategy. The core team led by Tim, consisted of 7 executives, including Randy (VP development), Joe (VP Sales and Marketing) and Jim (VP Strategy). It was supported by a total of 150 people, who provided input and ideas for a new corporate product direction. The following meeting was supposed to move the strategy into implementation. Max, the CEO of MyCorp led the meeting.

Max: OK, this is our final strategy meeting after one year of work. I'm proud that you all created this important plan for MyCorp's future, and I'm looking forward to its implementation.

Randy: Yes, it was hard work, and as a member of the core team I'd like to express my appreciation for the work done by all the people who helped us during the past 12 months.

Tim: OK, let's sign the final documents now, and look at the announcement text again before we implement the changes.

Randy: Not so fast! I need just one small clarification before we proceed: do I really need to give marketing the right to be involved in the development process right from the start? I have thought this over, and I think it might not be really necessary.

Joe: Wait a minute. This was a core element of our agreement. Without it I cannot agree to have development people in our pre-sales activities – they always promise too much anyway.

Randy: What? Without my people, even last year's miserable figures would not have been achieved.

Joe: Well, if we set off on this path again, the deal is off – you can count me out of the implementation.

Jim: OK, I had my doubts about whether this would work. I withdraw too.

Tim: I cannot believe this. We have worked 30 weekends over the past year to come to this agreement, and now you just want to blow it up – just because you have to give up some areas of your empires?

Randy / Joe both at the same time: This is not about empires; we just aren't ready yet. We need some clarification.

Tim: After one year's hard work, what else is there to clarify but your petty power games?

Max: umm, aah, well, let's close the meeting.

The meeting was unsuccessful, but outside in the hallway the CEO made the following comment:

Max: Tim, it seems I should have said something; I feel I let you and the company down.

Tim: You bet you did…

The strategic plan was never officially implemented. The separate and often competing sub-organisations continued to exist. But despite this obstacle, everybody followed and operated according to the ideas of the strategic plan. This demonstrates the power of a convincing idea, how it can survive despite being officially suppressed. Even the VPs followed the plan, although it was never publicly acknowledged.

What's beyond Innovation Cells?

Beyond innovation cells lies a vast continent of innovative possibilities. Although we understand innovation cells, there must be many more innovative possibilities we have not yet discovered. What is required to explore this treasure is primarily the open mind of the explorer himself. This mindset may be one of the most precious results of the work in an innovation cell. Viewed in this way, a new generation of professionals and leaders may emerge which is better suited to the challenging uncertainties of future innovations. Among the characteristics of this new leadership generation may be its ability to deal more effectively than ever before with power, and the cycles of creation and stability

The eternal power conflict

For every leader in any organisation, there comes a time when the diverging paths between the struggle of creating something with other people, versus the ability simply to tell others what they must do will become apparent. The former requires a mindset of true collaboration and an inner belief in human dignity. The latter requires the appropriate positional power and conformity to the system at hand. In the traditional system you are used to exercising positional power as a leader or manager. If you are a professional without positional power, you are used to conforming to demands.

From the manager or leader's point of view, the positional power represents the organisations trust in your capacity to "get the job done", whatever it may be. But can you really do this on your own in today's complex environment? Often you will realise that you have neither the in-depth understanding needed for the challenge at hand, nor the competence to structure a framework for a solution, so that the people working for you can just fill in the blanks. It is obvious that you need many talents to get the job done, one of which is the talent of leading or managing. But this talent alone is not enough, especially when innovation is the topic. In contrast to managing something which is known, such as marketing or production, innovation requires additional talents such as intuition, risk-taking or the ability to make technological assessments. A good solution becomes the result of true group collaboration: it is basically a co-created result.

This reality is in conflict with popular organisational belief. The manager is usually expected to be a hero figure. This belief in heroism is deeply ingrained in the formal structure of most organisations, in the form of reward systems, or the way people are appreciated and promoted. Because of this, you as a leader will always be tempted to go with the formal organisational flow. For the professional without positional power, the equally tempting attraction is to conform, because it allows you to shy away from taking responsibility. This is a deeply unsatisfactory situation. In innovation cells you are in a position to do something about it, whether you are a leader, a manager or a professional. Because of the innovation cell's autonomy, at least some individual and collective independence can be regained.

The eternal cycle of old and new

Innovation creates the truly new, something that did not exist before. But how do you know whether something is new? Let's consider how the old is like a backdrop against which the new can be recognised. Seen this way, the old and the new are tied together in an endless innovative cycle. Obviously any innovation will mature into something that at some point in the future will be considered old. It seems like an eternal cycle of decay and creation. The beauty of it is that in total it represents an expanding system. This reality points towards an abundance model, where innovation is the core energy for growth. Allow people and knowledge to interact, and innovation will happen.

The final piece of advice we can provide is to engage in an ongoing dialogue about what will sustain the growth of your company. To a degree it is a bit like gardening and farming: no gardener or farmer plants the same crops on the same fields or flowerbeds every year. Each year a good farmer will always ask "What shall I retain this year, and where should I try something new?" He will base his answer on a good understanding of the soil, the weather conditions, the markets and the resources at hand. The same sort of idea is true for leadership and management in today's business environment. You will continuously need to ask yourself "what will prolong my current products' life cycles for as long as possible?" And at the same time you need to ask "what will replace them as soon as possible?" Your innovative talents, in particular for disruptive innovations, combined with efficient and flexible production, are your keys to success. This creative force requires an ongoing dialogue from which innovation emerges. It is a never-ending quest. But while you are following it, you will be continually improving your questing skills.

- It was the leadership for innovation that had gone missing along the way.

□ *I wonder why...*

- Just start reading the book again from the beginning and you will remember. It happens to all of us!

The story of this book

This book itself was written using the principles of innovation cells. In this way the experience of writing the book is an example of the power of agile teams.

It all started with a bad experience. We had tried to write a book on the new innovative method by using the conventional team approach. After nearly two years we had to admit defeat. The story wouldn't flow, our enthusiasm in the subject matter just didn't come across, and those readers we did expose to the text, had to be resuscitated after collapsing with boredom! We had to admit defeat. The experience taught us to have the courage of our convictions, and use the principles of innovation cells to start afresh.

We were convinced that we had something worth sharing with the community of people who care about innovation, whether they are leaders, managers or other professionals. We decided to focus on what we had experienced ourselves. For various reasons, we also wanted to write the book as quickly as possible. Practising what we preached made a lot of sense.

Chronology

The book was written in just five months, from March to August 2004. During that time, two basic types of activities took place: five sessions of intensive work in the "Book" team room, and individual work done in the meantime between these sessions.

The team room

Following the rules of innovation cells we set up a "book project" team room in Uwe's private office. One wall in this room was used as a work area to create the overall structure of the book. On it we developed the structure of the book, its details and its key messages. The wall became a key resource of information and provided the backdrop for our work meetings.

Five months to write it

During the project we met up five times in the room for work sessions lasting between 1 and 3 days. During the sessions, the core of the book was written. We collated examples, structured the essays and worked out the overall flow structure. We had fun reading each other's texts, adjusted our styles and learned to appreciate each other's very different slants on the subject.

The process

The process of writing the book developed in part out of our conversations during the sessions. They were complemented by our own experience from large group work, such as Future Search conferences, and from our experience in innovation cells. From start to finish, we used ten different and distinct tasks to get from the first formulation of the idea of the book to the final document. The following describes these steps in more detail:

Mind-mapping

As in Future Search conferences, collective mind-mapping was made use of. One wall of our room was covered with a piece of paper roughly 1 metre by 4 metres. In the middle of this worksheet was written our central question "The new book?" We "mind-mapped" everything we felt was necessary for successfully writing the new book. With the aid of the map, we developed an understanding of what we thought it was most important to include in this book. For example, the layered structure of the book, with dialogue, essays and examples is a direct result of this activity.

Pictures

Using the next step, we wanted to get a clear picture, individually and as a team, of what were the important messages in the book. We decided first to draw pictures on cards, each depicting important messages.

Cross-interpretation

We then explained our individual pictures to each other. The person listening to the explanations wrote down his understanding of the message on the back of the card. The messages in written and in picture form turned out to be a great help throughout the project. We came up with 39 distinct themes we wanted to write about.

Flow

Talking about the pictures, it became clear that the themes could be grouped into 5 different but interconnected groups, which became the main chapters of the book. We used our wall again and posted the cards, sorted by key area, from left to right on top of the mind-map. After a bit of re-shuffling, we had in front of us the main flow of the book.

Examples

We then went back to the layers we had developed in the mind-map, and for each card asked the following question "What real life examples can we remember which support the message on the card?" We used Post-it notes to write down the working titles for the examples, and stuck them on top of the appropriate cards.

At this stage we decided to relate the examples, writing down as many as possible while we were still together in the team room. Nearly 80% of the examples were written in this way during our first meeting. It was a memory session, filling us with joy and sometimes sadness.

Essays

We finished our first session by asking the question "What are the subjects we want to write about in more detail, and where do we think we have original knowledge to share?" Again we used Post-it notes to identify topics, and attached them to the cards where they seemed to fit best. We called these parts of the book "essays".

At the end of our first work session, which lasted 3 whole days, we had developed together a clear understanding of what the book was going to be about. As well as its structure and key components, we had finished writing almost all the examples - and we felt somewhat washed-out!

We went our own ways, both having job and family to attend to, but continued to work individually, finishing the examples and writing the first drafts of essays and the main dialogue. We used digital photos of the wall as a "data bank" and backdrop for our individual work.

Main dialogue

Based on the work of the first session and the digital records of it, Burkard wrote a first draft of the author-reader dialogue, which became the backbone of the book. His bossy attitude came into its own here, but we managed to tone it down a little, making it more acceptable to readers!

Hyperlinks

Once the author-reader dialogue was in place, we worked on our individual essays and examples. In parallel, Uwe continued to use the "book wall" in his office, to work further on streamlining the structure of the book. In this way, both refined the structure of the book towards its final form, one using the author-reader dialogue, the other the "book wall".

The result of this phase was a first draft of the complete book. Only one more checkpoint meeting was required in between. Because at this time it was highly modularised (author-reader dialogue, essays and examples), the first attempt at integration was in the form of a hyper-linked document, which aligned the examples and essays within the five key areas of the book.

During one joint session, we positioned the examples and essays within the author-reader dialogue at text positions where they seemed to fit best. This was still a hyper-linked document. Based on this document we again worked individually, refining essays and examples, now in close relationship to the author-reader dialogue.

Checking

The fourth session in our team room was dedicated to checking whether what we had intended to write matched our actual draft of the book. This was a relatively easy task, because our "book wall" included all the details needed for the task. The cards and Post-it notes reflected themes, examples and essays, and as a whole, represented the flow of the book. We literally walked from left to right along this "book wall", looking first at all the Post-it notes, asking ourselves the question "Did we really do what we had posted?" If the answer was "Yes" we threw the post-it away. If it was "No", we thought what to do about it. After a decision was made we also threw these Post-its away.

Our next job was to look at the cards containing the pictures and texts. Again we asked, "Did we cover the message conveyed by the card?" If the answer was "Yes", it ended up in the waste-paper basket. If the answer was "No", we thought what do about it and decided who should take on the task. After this, all cards, Post-its and other memos on the wall had been removed, and the mind-map was all that remained.

This was the starting point for our final check. We looked at each comment on the mind-map, asking again "Have we met our objective?" We were pleasantly surprised to note that there were no outstanding items on the map.

Integration

Burkard integrated all the various parts of the book into one text document that could be used as the basis for printing, including formatting it to the publisher's requirement. When this was done, the unified document was handed over to Uwe, who picked up where Burkard had left off, doing his part editing and correcting. The document then went to Janet, who happily ironed out any linguistic creases, and finally the text was sent to the publisher.

Summary

Innovation Cell principles observed

This book was written, as described above, between March 10th and August 19th 2004. Although we spent a significant amount of time on it, the break-through points were the joint sessions. This time in total, was not more than 7 days, spread over 5 meetings. We are really quite pleased with the resulting book. We contribute our successful creation of the book to the fact that we worked as an innovation cell. The following key principles were particularly important: one room, a clear goal, motivated authors who knew and trusted each other, a deliberately short time-period for writing, and of course, the support of our families.

Personal experience

Uwe Weissflog:

The way we worked together makes the point about work structures that are simultaneously effective and enjoyable. The book emerged from our joint sessions, which were characterised by openness and trust. It is always amazing to me just what can be created, when the focus is on the power of creation. Writing this book makes the point about agile teams. Work can actually be very enjoyable and effective at the same time.

Burkard Wördenweber:

Not everyone is a born author. Most authors have to work hard to overcome inertia and remain motivated in order to achieve their goal. Here it was surprisingly easy to do as part of a dedicated team, and moreover the process of learning was rewarding. I personally had been able to encourage numerous engineers to create innovations, using innovation cells. Now I was glad to put myself through the same process, at the same time committing to paper what I had observed others doing.

Further reading

The books listed here allow the reader to explore further the core issues of innovation, in the context of the five perspectives portrayed in this book. The books are listed in the sequence of the chapters, and include a selection of authors, representing a wide spectrum of mainstream as well as unorthodox methods of making sense of the new, emerging business world. Titles in italics mark personal choices of the authors, which they particularly enjoyed and now recommend.

A growing awareness of being stuck

Collins, James C. und Porras, Jerry I. Built to Last, 1994, Harper Collins Publishers, New York

Cusumano, M. A., Nobeoka, K., 1998. Thinking beyond lean. The Free Press: New York

Deming, W., Edwards T., 1982, Out of Crisis, MIT, Cambridge

Drucker, Peter F., 1995, Management Challenges for the 21st Century, Truman Tally Books/ Dutton, Published by Penguin Group: New York

Hofstede, G., 1980, Culture's Consequences: International Differences in Work-Related Values, Beverly Hills, CA: Sage Publications

Hofstede, G., 2001, Lokales Denken, globales Handeln. Interkulturelle Zusammenarbeit und globales Management, Verlag Beck

Moore, Geoffrey A., 1995, Inside the Tornado, HarperCollins, New York

Treacy, Michael, Wiersema, Fred, 1995, The Discipline of Market leaders, Addison-Wesley publishing Company

Reaching a new perspective

Block, Peter, 1996, Stewardship – Choosing Service over Self-Interest, Berrett-Koehler Publishers, San Francisco

Csikszentmihalyi, M., 1994, Flow: The Psychology of Optimal Experience, Simon & Schuster, New York

Fromm, E., 1997, To Have or To Be?, Continuum, New York

Lewis, T., MD, Fari Amini, M.D., Richard Lennon, M.D., 2000, A General Theory of Love, Vintage Books, New York

Maslow, A.H., 1971. The Farther Reaches of Human Nature, Penguin Arkana

Maturana, Humberto R. Ph.D, Varela, Francisco J. Ph.D, 1998. The Tree of Knowledge, The Biological Roots of Human Understanding, Revised Edition, Shambala: Boston & London

Seligman, M.E.P, 1975, Helplessness: On depression, development and death, Freeman: San Francisco

Taking the plunge into the unknown

Gleick, J, 1987, Chaos, Making a new Science, Viking Penguin, New York

Morgan, Gareth, 1997. Images of Organization, Second Edition, Sage Publications

Owen, Harrision, 1997, Open Space Technology, Berrett-Koehler Publishers, San Francisco

Weisbord, Marvin R., Janoff, Sandra, 1995, Future Search, Berrett-Koehler Publishers, San Francisco

Wheatley, Margaret J., Kellner-Rogers, Myron, 1996. A simpler Way, Berret-Kohler Publishers, San Francisco

Wheatley, Margaret J., 1999. Leadership and the New Science, Discovering Order in a Chaotic World, Berret-Kohler Publishers, San Francisco

Change is hard work

Pinchot, G., 1985, Intrapreneuring: Why You Don't have to leave the Corporation to Become an Entrepreneur, New York 1985

Senge, Peter M., 1990. The Fifth Discipline: The Art and Practice of The Learning Organization. Curreny Doubleday

Senge, Peter M., Ross, Richard, Smith, Bryan, Roberts Charlotte, Kleiner, Art, 1994. The Fifth Discipline, Fieldbook: Strategies and Tools for Building a Learning Organization. Curreny Doubleday

Senge, Peter M., Ross, Richard, Smith, Bryan, Roberts Charlotte, Kleiner, Art, Roth, George. 1999. The Dance of Change, The Challenges to Sustaining Momentum in Learning Organizations: A Fifth Discipline Resource. Curreny Doubleday

Wördenweber, B., Wickord, W., 2003. Technology- und Innovations-management im Unternehmen, Methoden, Praxistipps und Softwaretools, 2nd and extended edition, Springer: Heidelberg

Selling innovations

Block, Z., MacMillan, I.C., 1993, Corporate Venturing: Creating New Businesses within the Firm, Boston
Mitchell, Stephen, 1988, Tao Te Ching, Harper and Row, New York

If all else fails, read:
Steven Fry, 1996, Making History, London;
Or, equally fascinating, but not quite so outrageously funny:
Robert Harris, 1993, Fatherland, Arrow;
and find out what could have happened, or how different the future might have been if history hadn't been quite so disruptive.

Acknowledgements

Our thanks go to Janet Wördenweber for English language amendments.

Acknowledgements for conscious and unconscious contributions are given to:

Adward F. Hart * Akihiro Nakamura * Alar Vasemägi * Albert F. Peter * Alexander Chong * Alexander Patt * Alexander Suhm * Alfred Katzenbach * Allan Behrens * Alyn Rockwood * Amparo Roca de Togores * Anderas Dengel * Andrea Dyer * Andreas Friedrich * Andreas Luttmann * Andreas Mueller * Andreas Pohl * Andreas Scheel * Andreas Silbernagel * Andreas Strobel * Andrew Fellows * Angelo DiViesto * Anja-Karina Pahl * Anne Stickley Michel * Anne-Liis Arulo * Antje Stroebe * Armin Vornberger * Arturas Kaklauskas * Axel Schröder * Barbara Stauder * Barry Heermann * Bas Possen * Ben Blemker * Ben Wang * Bernard Michael Gilroy * Bernd Ehrenberg * Bernd Pätzold * Bernhard Newe * Bernhard Newe * Bill Lee * Bill Mesley * Björn Abel * Bob Brown * Bob Whale * Bob Williams * Brad Holtz * Breytenbach Breyten * Britta Trompeter * Bror Salmelin * Bryan Stone * Burkhard Ott * C. Werner Dankwort * Carl Eitel * Carlo Capponi * Carlo Lombardi * Carol Scallan * Carolin Gebel * Carsten Neufeld * Cas Szczepanik * Charles Brassard * Charles T. Mazurie * Chris Gilbert * Christer Fernstrom * Christian Hart * Christian Pilz * Christian Schmidt * Christian Verstraete * Christine Frick * Christine Lingg-Thanner * Christine M. Merkel * Christoph Schneider * Christopher Bangle * Christopher Venning * Claudia Kadel * Cornelius Herstatt * Cornelius Neumann * Cristin Fitzpatrick Jameson * D. P. Dash * Dana Z. Anderson * Daniel Armbruster * David Whittaker * Deborah Cox * Deborah Dora * Detlef Decker * Detlef Korf * Deval M. Desai * Diane Coyle * Diane S. Menendez * Dianne Stuart * Dieter Bury * Dieter Rombach * Dieter Stauder * Dietmar Fischer * Dietmar Treichel * Dietmar Trippner * Dimitris Kiritsis * Dirk Bornwasser * Donald. B. Welbourn * Donna Blakemore * Doris Holland * Doris Holland * Dorothea Reese-Heim * Doug Finton * Doug Kreysar * Eberhard Kallenbach * Eberhard Zeeb * Eckard König * Eckart Miessner * Ed Dolan * Ed Grenda * Edgar H. Schein * Edward Klein * Edwin Courts * Egon Behr * Eike Tonismae * Ekkehart Frieling * Elke R. Janesh * Eric Hatch * Eric Scherer * Erich Lepiorz * Erik Sueberkrop * Ernst G. Schlechtendahl * Erwin Freudenreich * Erwin Marke * Federico R. Casci * Felix Cassellas * Florent Frederix * Florian Fischer * Florian Haake * Fozzy * Franca Giannini * Frances Pocock * Franco Naccari * Frank Kovacs * Frank Lillehagen * Frank Taschner * Frank-Lothar Krause * Frank-

Lothar Krause   *   Franz Feierabend   *   Franz Obinger   *   Gabi Sachs-Dücker   *
Gábor Renner   *   Gabriele M. Ganswindt   *   Gary Silverman   *   Gearold F. Johnson   *
Geoff Hatfield   *   Georg Kraft   *   Georg Rührnschopf   *   Georg Weissberger   *
Georges Zissis   *   Gérard Guilbert   *   Gerd Bierleutgeb   *   Gerd Herziger   *   Gerd
Schumann   *   Gerhard Reichinger   *   Gerhard Rotgeri   *   Gerold Häflinger   *   Gerry
Johnson   *   Gert Hildebrandt   *   Glen Allmendinger   *   Greg Spencer   *   Greg Spenzer
*   Guenter Harsy   *   Guenter Schweisshelm   *   Gunnar Bringfeldt   *   Gunnar Nilsson
*   Günter Reichart   *   Günter Warnecke   *   Gunter Zimmermeyer   *   Günther
Burghardt   *   Günther Sejkora   *   Gustav J. Olling   *   Guy Wills   *   H.-J. Meyer   *
Hanno Boekhoff   *   Hans de Jong   *   Hans Hagen   *   Hans-.Georg Frischkorn   *
Hans-Georg Metzler   *   Hans-Georg Stork   *   Hans-Joachim Schmidt-Clausen   *   Hans-
Joachim Stocker   *   Hans-Jürgen Linde   *   Hans-Kurt Luebberstedt   *   Hans-Ulrich
Knipps   *   Harald Briese   *   Hardi Nigulas   *   Harrison Owen   *   Hartmut Esslinger   *
Hector Fratty   *   Heike Findeis   *   Heiner Ruschmeyer   *   Heinrich Bunse   *   Heinz
Pfannschmidt   *   Heinz-Simon Keil   *   Helen Leemhuis   *   Helena Lindström   *
Helene Guillemin-Weiss   *   Helmut Kutt   *   Herbert Kircher   *   Herbert Leitner   *
Herbert Pietschmann   *   Hermann Riedel   *   Hermann Rutsch   *   Hideki Toyotama   *
Hiroyuki Hasegawa   *   Hitoshi Asakawa   *   Holm Tietz   *   Horst Dahm   *   Horst
Nowacki   *   Horst Nowacki   *   Horst Stevenson   *   Hubert Kals   *   Ip-shing Fan   *
Iris van Harpe   *   Isidro Laso   *   Italo Caroli   *   Jaan Parn   *   James Flaherty   *
James Kelly   *   Jan Andresoo   *   Jan Stanek   *   Jana Rae Brown   *   Jane A. Cash   *
Janet Jackson   *   Jarmo L. Suominen   *   Jeff Miller   *   Jeffrey Beeson   *   Jessica
Franken   *   Jian Jiao   *   Jim Brewer   *   Jim Rusk   *   Jim Stuart   *   Joachim Christ   *
Joachim Doering   *   Joachim Hauser   *   Joachim Rix   *   Joachim Schmitt   *   Joachim
Taiber   *   JoAnne Siebe   *   Joe Sullivan   *   Joel Orr   *   Joelle Lyons Everett   *   Joerg
Feldhusen   *   Joerg Sauer   *   Johan W.A.M. Alferdinck   *   John Bradbrook   *   John
Gach   *   John Veasey   *   John Zurick   *   Jordi Portella   *   Jörg Wallaschek   *   Jose
Antonio Estevez   *   Josef Schug   *   Judi Martindale   *   Judi Neal   *   Juergen Sellentin
*   Juergen Weber   *   Juha Halonen   *   Julia L. Ritchie   *   Julian Vincent   *   Jürgen
Antonitsch   *   Jürgen Blum   *   Jürgen Gausemaier   *   Jürgen Hille   *   Jürgen Plato   *
Kai Storjohann   *   Kare Rumar   *   Karen R. Dickinson   *   Karl Dokter   *   Karl
Heinrich Oppenländer   *   Karsten Eichhorn   *   Karsten Rose   *   Kathlyn Hall   *   Ken
Alexander   *   Ken Homer   *   Kevin Marschall   *   Kim Viborg Andersen   *   Klaus Di-
eter Thoben   *   Klaus G. Saul   *   Klaus Pasemann   *   Klaus Schäfer   *   Knuth Götz   *
Knuth Schmidt   *   Koji Oe   *   Konrad Wegener   *   Kristen Buckley   *   Kristina Enden
*   Kulle Tarnov   *   Kurt Gelfgren   *   Kurt Neuffer   *   Kvisuri Tikkanen   *   Lee Robie
*   Leif Edvinsson   *   Lennart Dellby   *   Leonard Livschitz   *   Liane Brockort   *
Linda Brentano   *   Lora Hein   *   Lorne Karasik   *   Lothar Heggmair   *   Ludwig Lingg
*   Lukas Schwenkschuster   *   Lutz Blencke   *   Lutz Völkerath   *   Lynne Palazzolo   *
Maggie Rogers   *   Malle Schaffrik   *   Manfred Sammet   *   Manfred Spitzer   *
Manuela Gyr   *   Marco Taisch   *   Marie Redmond   *   Marjo Kauhaniemi   *   Mark R.
Jones   *   Mark S. Rea   *   Mark Valdma   *   Mark Withers   *   Markrab Ingrid   *   Mar-
kus Baum   *   Markus Köhl   *   Markus Schichtel   *   Markus Schmitt   *   Martin Bern-
hard   *   Martin Eigner   *   Martin Enders   *   Martin Kohler   *   Martin Leith   *   Martin
Neads   *   Martin Rheinbach   *   Martin Stark   *   Martin Watzlawek   *   Martina
Schmidt   *   Marv Weisbord   *   Mary L. Azzolini   *   Masanori Sato   *   Massimo Sig-
nani   *   Mathias Kothe   *   Mati Ruul   *   Matti Tähkäpää   *   Max Dorando   *   Max
Lemke   *   Max Neumeier   *   Michael C. Holt   *   Michael Hartinger   *   Michael Her-
man   *   Michael Klinge   *   Michael Kramer   *   Michael Lippmann   *   Michael Moser
*   Michael Pohle   *   Michael Sivak   *   Michael Theobald   *   Michael Wieschalka   *
Michel Schreyer   *   Mike Evans   *   Mitsuo Yamada   *   Monika Bartelt   *   Muriel

Kennedy * Muriel Pigatti * Nancy Grengs * Nico Correns * Nicole Arthaud *
Nikolaus Ondracek * Nikolaus Risch * Nils Bergheim * Nils Labahn * Nils
Porrmann * Noele C. Williams * Norbert Rech * Norbert Thom * Norihito Ya-
maguchi * Noriko Tsuchiya * Olavi Luotonen * Oskar Gelinek * Oskar Klemisch
* Palat Sreevalsan * Pam Winkler * Pat Marchant * Pat Murphy * Paul Chesh-
ire * Paul Hearn * Paul Hearn * Paul Mulvanny * Paul Sawyer * Penny Robie
* Peter Dadam * Peter Gumbsch * Peter Hanke * Peter Lagies * Peter Merschen
* Peter Moritz Delwing * Peter R. Boyce * Peter Roche * Peter Sachsenmeier *
Peter Schwarberg * Peter Thorne * Piero Perlo * Rainer Bugow * Rainer
Gödecker * Rainer Israel * Rainer Kauschke * Rainer Weinzettl * Ralf China *
Rama Camara * Rauli Sorvari * Reiner Anderl * Reinhard Bertram * Reinhard
Schmitt * Renate Kerbst * Ricardo Goncalves * Richard Baumann * Richard
Häusler * Richard Janse van Rensburg * Richard L. Shoemaker * Ritva Parhamaa
* Robert C. Buxbaum * Robert D. Scallan * Robert Goodsell * Roger Breisch *
Roland Blomer * Roland Lachmayer * Roman Tomassin * Ron Edwards *
Ronald Mackay * Ronald Sander * Rüdiger Koch * Rüdiger Singer * Sabina
Battaglino * Sabine Kugel * Sally Theilacker * Samantha Amit * Sari Jouhki *
Sarita * Satu Vainio * Sayuri Suzuki * Sebastian März * Serge Wynen * Ser-
gio Gusmeroli * Shinya Omori * Siegfried Heinz * Siegfried Reiniger * Silvia
Ansaldi * Silvia Specht * Simone Schlegel * Stacy Flaherty * Stefan Jähnichen
* Stephan Trippe * Stephan Völker * Steve Hinkle * Steve Widdet * Sue Miller
Hurst * Susan Smyth * Susanne Weber * Susanne Weber * Takeshi Kimura *
Tamas Varady * Teresa de Martino * Terry Davidson * Terry Kennair * Thomas
Barth * Thomas Börnchen * Thomas Fischer * Thomas J. Hurley * Thomas
Lehnert * Thomas Neumann * Tonguc Ünlüyurt * Trac Tang * Tsutomu Yama-
moto * Uli Dombrowsky * Ulrich Ahle * Ulrich Gehrke * Ulrich Hilleringmann
* Ulrich Kramer * Ulrich Rückert * Ulrich Saelzer * Ulrike Guelich * Ulrike
Langer * Ute John * Uwe Gühl * Uwe Kaufmann * Uwe Kleinheidt * Uwe
Müller * Uwe Ottersbach * Valerie Baum Lingeman * Veronica Rolle Green *
Veronika Hrdliczka * W.A. Geist * Walter Ziegler * Warner B. Wims * Werner
Popp * Werner Rixen * Wilfried Rostek * Wilhelm Brandenburg * Wilhelm
Horstmann * Wilhelm Kerschbaum * Wilhelm Schäfer * Will Verity * William
Kimberley * Willibert Schleuter * Winfried J. Huppmann * Wiro Wickord *
Wolfgang Barth * Wolfgang Geist * Wolfgang Heyn * Wolfgang Huhn * Wolf-
gang Kurz * Wolfgang Sassin * Yann Barbaux * Yasmina Bock * Yoshirou Na-
gamura * Yumi Hoshi * Zoe Nicholson *

Druck: Krips bv, Meppel
Verarbeitung: Litges & Dopf, Heppenheim